Exposure Assessment and Health Effects Due to Ionizing and Non-Ionizing Radiation

This book describes and outlines exposure scenarios of ionizing and non-ionizing radiation in occupational and residential settings within the African continent, where technological developments and mining activities are the main sources of radiation exposure. To better understand these exposure scenarios, this book focusses on all exposure aspects of both ionizing and non-ionizing radiation from an African perspective.

This book

- Covers occupational and residential exposure to ionizing radiation emanating from mining activities, particularly from an African perspective
- Explores exposure to radon and uranium from abundant mines and tailings
- Reviews non-ionizing radiation exposure assessments for communication and power sources
- Discusses health effects of ionizing and non-ionizing radiation among the general public
- Examines health effects of exposure to mining, electric utility, and communication device operations

This book is aimed at graduate students and researchers in bioengineering and environmental health, and radiobiology.

Exposure Assessment and Health Effects Due to Ionizing and Non-Ionizing Radiation

An African Perspective

Phoka Caiphus Rathebe

CRC Press
Taylor & Francis Group
Boca Raton London New York

CRC Press is an imprint of the
Taylor & Francis Group, an **informa** business

First edition published 2026
by CRC Press
2385 NW Executive Center Drive, Suite 320, Boca Raton FL 33431

and by CRC Press
4 Park Square, Milton Park, Abingdon, Oxon, OX14 4RN

CRC Press is an imprint of Taylor & Francis Group, LLC

ISBN: 9781032812540 (hbk)
ISBN: 9781032832722 (pbk)
ISBN: 9781003508588 (ebk)

DOI: 10.1201/9781003508588

Typeset in Times
by codeMantra

Contents

Preface

Radiation exposure, whether from natural or anthropogenic sources, poses significant health risks globally. In Africa, however, the assessment of ionizing and non-ionizing radiation exposure remains underexplored, despite increasing industrialization, urbanization, and technological advancement. This book aims to bridge that knowledge gap by providing a comprehensive analysis of radiation exposure and its health effects from an African perspective.

The motivation behind this work stems from the growing concern over environmental and occupational radiation hazards in African communities. While ionizing radiation from sources such as radon, medical imaging, and mining activities has been extensively studied in other regions, research in Africa is often fragmented. Similarly, non-ionizing radiation, including ultraviolet (UV) rays and electromagnetic fields from mobile communication technologies, has received limited attention, even as the use of these technologies surges across the continent.

This book is structured to provide a holistic understanding of radiation exposure assessment, covering key topics such as measurement techniques, risk evaluation, and mitigation strategies. It also highlights case studies from African countries, illustrating real-world scenarios and the unique challenges faced by researchers, policymakers, and public health officials.

By presenting an African-centred approach, this book aims to contribute to global discussions on radiation safety while advocating for improved monitoring, regulation, and public awareness. It is intended for researchers, environmental health professionals, students, and policymakers seeking a deeper understanding of radiation exposure and its health effects in the African context.

I hope this work inspires further research and policy development to safeguard public health and promote sustainable development in Africa.

Phoka C. Rathebe, PhD

Author Bio

Phoka Caiphus Rathebe is an emerging scientist and an NRF Y-rated scholar with a strong research focus on environmental and occupational health, and radiation exposure assessments. He is currently an Associate Professor in the Department of Environmental Health at the University of Johannesburg. Rathebe holds multiple academic qualifications in Environmental Health from the Central University of Technology, Free State, including a National Diploma (cum laude, 2013), a BTech (cum laude, 2014), a Master of Health Sciences (cum laude, 2017), and a PhD (2021). His expertise extends beyond academia, as he actively reviews environmental and occupational health manuscripts for international peer-reviewed journals, IOP, and IEEE conferences.

As a dedicated researcher and supervisor, Rathebe mentors postgraduate students working on projects in occupational and environmental health. Since 2018, he has published over 60 manuscripts in peer-reviewed journals and conference proceedings, contributing valuable insights to the field. His work continues to shape the discourse on radiation exposure and public health, particularly within the African context.

1 Ionizing radiation exposure from various sources and activities in Africa

An overview

1.1 BACKGROUND OF THE CHAPTER

The pervasive influence of ionizing radiation in our modern world demands a comprehensive understanding of its sources, manifestations, and potential impacts on human health and the environment. Ionizing radiation, generated from both natural and anthropogenic sources, permeates various aspects of daily human lives, playing a crucial role in fields such as medicine, industry, and energy production. As we navigate an era marked by unprecedented technological advancements and increasing reliance on nuclear technologies, a nuanced exploration of ionizing radiation becomes imperative for students, researchers, and professionals alike. This book chapter delves into the multifaceted realm of ionizing radiation, offering a systematic examination of its diverse sources and activities. From cosmic rays and terrestrial radionuclides to medical diagnostic procedures and industrial applications, this chapter aims to provide a holistic overview of the sources contributing to human exposure. Understanding the complexities of ionizing radiation is not merely an academic pursuit; it is a prerequisite for informed decision-making, exposure assessment, risk characterization, and the development of effective safety measures.

The chapter will navigate through the historical evolution of our engagement with ionizing radiation, tracing its trajectory from scientific curiosity to essential tools in medical diagnosis, industrial processes, and energy generation. Moreover, it will scrutinize the challenges associated with ionizing radiation, including its potential health risks, environmental implications, and the regulatory frameworks established to mitigate these concerns. As educators, researchers, and practitioners, our responsibility extends beyond imparting theoretical knowledge; we must equip the next generation with the skills to navigate the complexities of ionizing radiation responsibly. By fostering an in-depth understanding of its sources and activities, this chapter aims to contribute to the development of a knowledgeable and conscientious community capable of harnessing the benefits of ionizing radiation while safeguarding the well-being of current and future generations.

DOI: 10.1201/9781003508588-1

1.2 KNOWN SOURCES OF RADIATION EXPOSURE IN AFRICA

1.2.1 Radon from Mining

Mining activities, particularly those associated with the extraction of uranium, gold, and other radioactive minerals, can contribute significantly to elevated radon levels in the immediate environments and those proximal to mines (Moshupya et al. 2019). Radon, a colourless and odourless radioactive gas, is a natural byproduct of the decay of uranium and thorium in soil and rocks (Varshney 2013). Exposure to radon from mining sectors in Africa presents a significant occupational health concern due to the continent's extensive mining activities, particularly in regions rich in uranium deposits (McCormack and Schüz 2012). Workers in mines, especially those involved in uranium mining, are at increased risk of radon exposure, which can lead to chronic respiratory diseases such as lung cancer (Semenova et al. 2020). The lack of comprehensive scientific literature in understanding risks identification, assessments, and management exacerbates the risks faced by miners, with inadequate ventilation systems and safety measures further contributing to elevated radon levels in mining environments (Vance et al. 2014). Moreover, informal mining practices prevalent in many African countries pose additional challenges in monitoring and mitigating radon exposure among workers. Efforts to address this issue require collaborative initiatives between governments, mining companies, and relevant stakeholders to implement stringent safety standards, enhance risk-monitoring protocols, and provide comprehensive training programmes to protect worker health and well-being. Research efforts should focus on assessing radon levels in different mining settings, identifying effective mitigation strategies, and raising awareness about the risks associated with radon exposure in the mining sector (Mudd 2008). Through concerted action and investment in occupational health measures, the adverse effects of radon exposure in African mining sectors can be mitigated.

1.2.2 Uranium from Mining and Power Plants

The mining and utilization of uranium in nuclear power plants are pivotal to the global energy landscape. While nuclear energy offers a low-carbon alternative, the extraction and processing of uranium present environmental and health challenges (Krūmiņš and Kļaviņš 2023). Mining activities release uranium into the surrounding environment, potentially contaminating water sources and affecting ecosystems (Yi et al. 2020). In nuclear power plants, proper safety measures, waste management, and adherence to regulatory requirements are crucial to prevent radioactive releases (Raj et al. 2006). The discussion surrounding uranium emphasizes the importance of balancing energy needs with environmental and public health considerations.

1.2.3 X-rays from Healthcare Sectors

X-rays, a form of ionizing radiation, play a vital role in medical diagnostics, enabling healthcare professionals to visualize internal structures for diagnosis and treatment (Hussain et al. 2022). While X-rays are indispensable in healthcare, the associated ionizing radiation raises concerns about potential health risks. It is imperative for healthcare sectors to implement strict protocols for the justified use of X-rays, ensuring that exposure levels are minimized while maintaining diagnostic efficacy (Code 2024). Despite their importance, challenges such as limited access to equipment, shortage of trained personnel, and inadequate infrastructure persist in many African countries, hindering the widespread availability and quality of X-ray services (Jiménez et al. 2014). Additionally, concerns regarding radiation safety and dose management underscore the need for vigorous regulatory frameworks and training programmes to ensure the safe and effective use of X-rays in healthcare settings. Efforts to improve access to X-ray services in Africa have been underway, with initiatives focusing on expanding infrastructure, enhancing training programmes, and promoting technology transfer (Suleiman and Muhammad 2022). However, disparities in access to healthcare resources and funding constraints continue to pose significant barriers to the equitable provision of X-ray services across the continent (Signé 2021). Furthermore, the emergence of new imaging modalities and advancements in technology present both opportunities and challenges for the future of X-ray diagnostics in Africa (Olawole et al. 2024). Research efforts should prioritize addressing the unique healthcare needs and resource limitations in Africa, exploring innovative strategies to optimize the use of X-rays, improve patient outcomes, and reduce workers' exposure.

1.2.4 Ultraviolet (UV) exposure from the sun

UV radiation from the sun is a natural component of sunlight and is essential for various biological processes, including vitamin D synthesis. However, overexposure to UV radiation poses significant health risks, such as skin cancer, premature ageing, and eye damage (Neale et al. 2023). Balancing the benefits and risks of sun exposure requires public awareness campaigns, promotion of sun-safe behaviours, and the development of protective measures such as sunscreen use, protective clothing, and avoiding peak sunlight hours (Hadjee 2020). Understanding the importance of sun safety is crucial for promoting lifelong habits that minimize the adverse effects of UV radiation.

1.3 IONIZING RADIATION OCCUPATIONAL EXPOSURES FROM MINING, HEALTH, UV, AND POWER GENERATION ACTIVITIES IN AFRICA

This section addresses the multifaceted challenges and considerations associated with ionizing radiation across various occupational sectors on the African continent. The discussion encompasses the occupational exposure risks faced by workers involved in mining activities, particularly those related to uranium and gold extraction, X-rays in the healthcare sector, and UV radiation, primarily from sun exposure among outdoor workers. Furthermore, this section highlights the importance of environmental and occupational safety considerations for ionizing radiation in power generation plants.

1.3.1 Occupational radon exposure

Occupational radon exposure in Africa presents a multifaceted challenge that necessitates thorough examination. Despite being a continent rich in natural resources, Africa faces significant risks associated with radon exposure in various occupational settings, particularly in mining, construction, and manufacturing industries (Voldřich 2019). The prevalence of mining activities across Africa, coupled with inadequate regulatory frameworks and limited awareness, exacerbates the risk of radon exposure among workers. Furthermore, the informal nature of many workplaces in Africa poses additional challenges in implementing effective radon mitigation measures and ensuring occupational safety standards. Addressing occupational radon exposure in Africa requires a comprehensive approach that integrates awareness campaigns, regulatory enforcement, and technological advancements in radon monitoring and mitigation (Schoonhoven 2012). Collaborative efforts between governments, industries, and academic institutions are crucial in developing and implementing effective policies and interventions to protect workers from radon-induced health risks. Additionally, capacity-building initiatives aimed at enhancing radon awareness and monitoring capabilities among occupational health professionals are essential for mitigating the health impacts of radon exposure in African workplaces.

Moreover, research efforts focusing on occupational radon exposure in Africa should prioritize understanding the specific contextual factors influencing exposure levels and health outcomes among workers (Schoonhoven 2012). This includes assessing the effectiveness of existing mitigation measures, exploring innovative technologies for radon monitoring, and identifying vulnerable occupational groups at higher risk of radon exposure. By addressing these challenges and leveraging collaborative approaches, African industries can mitigate the adverse effects of occupational radon exposure and safeguard the health and well-being of their workforce.

1.3.2 Occupational uranium exposure in mining and power plants

Occupational uranium exposure in African mining and power plants presents a pressing concern, with workers facing heightened risks of radiation-related health issues (Hecht 2009). The continent's abundant uranium resources drive extensive mining operations, while nuclear power plants contribute significantly to electricity generation (DeBoom 2020). However, inadequate regulatory frameworks and enforcement mechanisms exacerbate the risks faced by workers in these environments. The extraction, processing, and utilization of uranium expose workers to various hazards, including radiation-induced illnesses such as cancer and respiratory diseases (Golden et al. 2021). Urgent measures are needed to address this occupational health challenge comprehensively. To develop healthier mining and power plants, research efforts should prioritize several key areas. First, there is a significant need for enhanced monitoring and assessment of occupational uranium exposure levels in mining and power plant settings across Africa. Second, comprehensive studies are required to better understand the health impacts of long-term uranium exposure on workers, including the prevalence of radiation-related diseases and associated risk factors. Furthermore, research should focus on developing and implementing effective mitigation strategies to reduce uranium exposure and promote occupational safety in these industries.

1.3.3 Occupation UV exposure

UV exposure from the sun poses significant occupational health risks among workers in Africa, particularly those engaged in outdoor activities such as agriculture, construction, and mining (Linde et al. 2022). The continent's geographic location near the equator results in high levels of UV radiation, exacerbated by factors such as altitude and atmospheric conditions (Roro et al. 2016). Prolonged exposure to UV radiation can lead to various health issues, including sunburn, skin cancer, cataracts, and immune suppression (Neale et al. 2023). Many workers in Africa lack access to adequate sun protection measures, such as protective clothing, hats, and sunscreen, further increasing their vulnerability to UV-related health problems (Hadjee 2020). The impact of UV exposure is compounded by socioeconomic factors, with workers in low-income settings often facing greater risks due to limited access to healthcare and educational resources (Passeron et al. 2023). Additionally, cultural norms and practices may discourage the use of protective measures, further exacerbating the risk of UV-related health issues among workers (Julian 2017). Efforts to address UV exposure in workplaces require raising awareness about the importance of sun safety, providing training on sun protection measures, and implementing policies to ensure the availability of protective equipment.

1.3.4 Occupational exposure to X-rays

Occupational exposure to X-rays presents a significant health risk in various sectors, particularly in healthcare, where workers routinely handle and work with X-ray equipment for diagnostic and therapeutic purposes (Albander 2021). While X-rays are invaluable tools for medical imaging, prolonged or excessive exposure can lead to adverse health effects, including radiation-induced cancers and genetic mutations (Oakley et al. 2019). Healthcare professionals such as radiographers, radiologists, and technicians are particularly at increased risk due to their frequent interaction with X-ray machines (Sharma et al. 2019). To mitigate occupational exposure to X-rays, implementation of stringent safety protocols, including the use of protective gear such as lead aprons, thyroid shields, and dosimeters to monitor radiation levels, is required (Othman et al. 2023). Regular training programmes on radiation safety and awareness campaigns are essential to educate healthcare workers about the risks associated with X-ray exposure and the importance of adherence to safety protocols (Sarman et al. 2016).

Advancements in technology have led to the development of safer X-ray equipment with reduced radiation doses, contributing to the overall reduction of occupational exposure risks (Johnston et al. 2010). However, challenges persist, especially in resource-limited settings and continents, like Africa, where access to modern equipment and training opportunities may be limited. Research in this area focuses on assessing radiation exposure levels among healthcare workers (Gbetchedji et al. 2020), identifying occupational hazards, and evaluating the effectiveness of radiation protection measures (Lewis et al. 2023). More studies are needed to inform evidence-based guidelines and policies aimed at minimizing occupational exposure risks and ensuring the safety and well-being of healthcare workers in environments where X-ray equipment is utilized (Gbetchedji et al. 2020).

1.4 METHODOLOGIES TO ASSESS OCCUPATIONAL IONIZING RADIATION EXPOSURE

1.4.1 Radon Personal Exposure Assessments

Radon personal exposure assessments are significant for understanding individual risks and implementing targeted interventions to mitigate radon-related health hazards, particularly in mining environments and households proximal to mines (Mphaga et al. 2024). Radon, a naturally occurring radioactive gas, poses significant health risks when inhaled, including an increased risk of lung cancer (Kholopo and Rathebe 2024). Therefore, accurate assessment methodologies are essential to inform risk management strategies and protect the respiratory health of individuals exposed to elevated radon levels (Mphaga et al. 2024). In mining settings, personal radon exposure assessments involve equipping workers with wearable dosimeters or personal radon monitors to measure radon concentrations directly in their breathing zone (Ahmad et al. 2021). These devices provide real-time or integrated measurements of radon exposure, allowing for the identification of high-exposure tasks or work areas within the mine (Ahmad et al. 2021). Additionally, biological markers such as radon decay products in urine or exhaled breath may be analysed to assess internal radon exposure among miners (Linhares et al. 2017).

Biological markers offer a promising avenue for assessing internal radon exposure among miners (Robertson et al. 2013). This internal health risk assessment strategy could be worthy implemented among African miners. Radon decay products, such as polonium214 and lead214, can be measured in biological samples such as urine or exhaled breath (Dunn et al. 2003). These markers serve as indicators of recent radon exposure and provide valuable insights into the internal dose received by miners. Analysing radon decay products in biological samples allows for the quantification of radon intake and absorption rates, aiding in the estimation of individual exposure levels. Furthermore, biological monitoring enables the assessment of variations in radon exposure over time and across different work environments within the mine. Integrating biological markers with traditional exposure assessment methods enhances the accuracy and comprehensiveness of radon exposure monitoring programmes in mining settings (Van Schalkwyk 2005). However, challenges such as background levels of radon decay products and variability in individual metabolism must be addressed to ensure the reliability and validity of biological monitoring data (Makumbi et al. 2024).

Similarly, in households proximal to mines, personal radon exposure assessments play a crucial role in characterizing individual risks and informing targeted interventions (Mphaga et al. 2024). Residents may wear personal radon dosimeters or carry passive radon detectors to measure their radon exposure levels throughout the day (Tamakuma et al. 2020). These measurements provide valuable insights into variations in radon concentrations within the home and help identify activities or locations associated with higher radon exposure (Tamakuma et al. 2020). Advancements in wearable sensor technology and biomonitoring techniques have facilitated the development of innovative personal radon exposure assessment tools, enhancing the accuracy and efficiency of measurements (Lin et al. 2022). These tools enable researchers to quantify individual radon exposures more effectively and tailor interventions to specific needs. Despite significant progress in personal radon exposure assessment methodologies, challenges remain, including the need for standardized protocols, calibration procedures, and data interpretation guidelines (Kholopo and Rathebe 2024). There is a need for improved personal radon exposure assessment techniques, ultimately contributing to the protection of human health in mining environments and communities proximal to mines.

1.4.2 Uranium personal exposure assessments

Uranium personal exposure assessments in Africa are paramount due to the prevalence of uranium mining activities across the continent. These assessments involve the measurement and analysis of uranium concentrations in various environmental media and biological samples to evaluate individual exposure levels (Guéguen et al. 2017). Methods for uranium personal exposure assessments include wearable dosimeters, passive samplers, and biomonitoring techniques. These tools enable the quantification of uranium intake through inhalation, ingestion, and dermal contact routes (Papadopoulou et al. 2020). Additionally, biological markers such as uranium isotopes in urine or blood samples provide valuable insights into internal exposure levels among individuals (Grison et al. 2016). Uranium exposure assessments are particularly crucial for workers in uranium mining and processing facilities, where occupational exposures may be elevated.

In addition to occupational settings, residents living near uranium mines or processing plants may also be at risk of elevated uranium exposure levels. Therefore, comprehensive uranium exposure assessments encompass both occupational and environmental exposures. Challenges in uranium exposure assessments include variability in exposure pathways, background levels of uranium in the environment, and the need for standardized protocols and quality assurance measures (Rathod et al. 2023). Collaborative efforts are essential to address these challenges and improve the accuracy and reliability of uranium personal exposure assessments in Africa. By implementing robust risk assessment methodologies and monitoring programmes, health risks associated with uranium exposure and protection of the well-being of workers and communities across the continent can be effectively mitigated.

1.4.3 UV personal exposure assessments

UV radiation, emitted by the sun, is a ubiquitous environmental factor with both beneficial and harmful effects on human health. In Africa, where sunlight exposure is abundant, understanding UV person0al exposure is crucial due to its implications for skin cancer, eye damage, and other health outcomes. Exposure to UV radiation is associated with various health effects, including sunburn, premature ageing of the skin, and an increased risk of skin cancer, particularly melanoma (D'Orazio et al. 2013). Additionally, UV exposure can cause eye-induced health conditions such as cataracts and photokeratitis (Thomas et al. 2012). In Africa, where the majority of the population resides in regions with high levels of UV radiation, individuals are particularly vulnerable to these health risks (Lucas et al. 2010).

Several methods are used to assess UV personal exposure, ranging from subjective self-reporting to objective measurement techniques. Self-reporting methods, such as questionnaires and diaries, rely on individuals' recall of their sun exposure behaviour and protective measures (Dobbinson et al. 2014). While these methods are cost-effective and easy to implement, they may be subject to recall bias and inaccuracies (Parr 2008). Objective measurement techniques, such as wearable UV dosimeters and UV cameras, provide quantitative data on UV exposure levels (Huang and Chalmers 2021). Wearable dosimeters, typically worn on the wrist or clothing, measure UV radiation intensity in real-time and record cumulative exposure over time. UV cameras capture images of UV radiation distribution on the skin's surface, allowing for spatial analysis of exposure patterns (Turner et al. 2020). These objective methods offer valuable insights into individual UV exposure levels and behaviour, facilitating more accurate risk assessments.

Despite the importance of UV personal exposure assessment, several challenges exist in implementing these methods in Africa. Limited access to advanced UV measurement technologies, such as wearable dosimeters and UV cameras, may hinder comprehensive exposure assessments (Huang and Chalmers 2021). Additionally, socioeconomic factors, including poverty and lack of awareness, may contribute to suboptimal sun protection behaviour and increase vulnerability to UV-related health risks (Miller et al. 2017). Environmental factors such as cloud cover and atmospheric pollution can affect UV radiation levels, complicating exposure assessments (Bais et al. 2018). In rural areas, where access to healthcare and sun protection resources may be limited, individuals may face higher levels of UV exposure without adequate protection measures. Addressing the challenges of UV personal exposure assessment in Africa requires a multidisciplinary approach. Investment in UV measurement technologies and infrastructure, including the dissemination of wearable dosimeters and UV cameras, can enhance the capacity for accurate exposure assessments. Furthermore, integrating UV exposure data into public health surveillance systems can inform policy decisions and interventions aimed at reducing UV-related health burdens.

1.4.4 X-ray Personal Exposure Assessments

In Africa, where access to healthcare and workplace safety standards varies widely, ensuring accurate assessment of X-ray personal exposure is essential for protecting the health and safety of workers and patients. X-rays are a form of ionizing radiation used in medical imaging, non-destructive testing, and other applications (Hanke et al. 2008). While X-rays are valuable tools for diagnosing medical conditions and detecting defects in materials, repeated or high-dose exposure can lead to adverse health effects, including tissue damage, radiation burns, and an increased risk of cancer (Chaturvedi and Jain 2019). In healthcare settings, radiographers, radiologists, and other healthcare workers are at risk of occupational X-ray exposure, while patients undergoing diagnostic procedures may also face exposure risks (Wallin et al. 2019). Several methods are used to assess X-ray personal exposure, ranging from direct measurement techniques to indirect dosimetry approaches. Direct measurement methods involve placing dosimeters, such as film badges or thermoluminescent dosimeters (TLDs), on individuals to measure the radiation dose received (Dube 2015). These dosimeters record the cumulative dose over a specific period, providing quantitative data on X-ray exposure levels.

Indirect dosimetry approaches, such as environmental monitoring and computational modelling (Tootell et al. 2014), estimate X-ray exposure based on factors such as equipment settings, procedure duration, and distance from radiation sources. While indirect methods are less invasive than direct measurement techniques, they may be subject to uncertainties and assumptions, particularly in complex healthcare environments with multiple sources of radiation (Meghzifene et al. 2010). Several challenges exist in implementing these methods in Africa. Limited access to dosimetry equipment and resources, as well as inadequate training in radiation safety, may hinder accurate exposure monitoring in healthcare facilities and industrial settings. Additionally, the lack of standardized protocols and guidelines for X-ray dosimetry in Africa can lead to inconsistencies in exposure assessments and risk management practices. Furthermore, cultural beliefs and misconceptions about radiation may contribute to reluctance among healthcare workers and patients to use protective measures, such as lead aprons and shielding devices, during X-ray procedures. In some cases, economic constraints and resource limitations may prevent healthcare facilities from implementing proper radiation safety protocols, putting both workers and patients at risk of overexposure. Standardizing dosimetry protocols and guidelines for X-ray exposure assessment is essential for ensuring consistency and accuracy in exposure monitoring across healthcare facilities and industrial settings. Regulatory frameworks should be strengthened to enforce compliance with radiation safety standards and promote the adoption of best practices in X-ray dosimetry and risk management.

1.5 CONCLUSION

In Africa, radiation exposure arises from various sources, including occupational activities in mining, healthcare, UV exposure from the sun, and power generation. These sources contribute to the continent's overall radiation burden, necessitating effective methodologies for assessing and managing occupational ionizing radiation exposure. Mining activities in Africa are a significant source of ionizing radiation exposure, particularly due to the presence of uranium and other radioactive elements in mineral deposits. Workers in mines are at risk of exposure to radon gas and its decay products, which can lead to lung cancer and other respiratory diseases. Health and safety measures, along with regular monitoring using dosimetry, are essential to mitigate these risks. In healthcare settings, medical personnel are exposed to ionizing radiation during diagnostic and therapeutic procedures. Proper training, use of protective equipment, and adherence to safety protocols are crucial in minimizing occupational exposure. Dosimetry plays a vital role in monitoring radiation doses received by healthcare workers and ensuring compliance with safety standards.

UV exposure from the sun is another significant source of radiation exposure in Africa, particularly for outdoor workers. Prolonged exposure to UV radiation can lead to skin cancer and other health issues. Effective sun protection measures, such as wearing protective clothing and using sunscreen, are essential for reducing UV exposure among outdoor workers. Power generation activities, including nuclear power plants, also contribute to ionizing radiation exposure in Africa. Proper management of radioactive waste and strict adherence to safety regulations are critical in minimizing the risk of radiation exposure among workers in these facilities. Methodologies for assessing occupational ionizing radiation exposure in Africa vary depending on the source of exposure. Dosimetry, both external and internal, is commonly used to measure radiation doses received by workers. Biological monitoring, such as analysing radon decay products in urine or exhaled breath, can provide valuable information on internal radon exposure among miners. Effective management of radiation exposure in Africa requires a multifaceted approach, including robust regulatory frameworks, comprehensive training programmes, and the use of advanced dosimetry techniques. By implementing these measures, African countries can ensure the safety and well-being of workers exposed to ionizing radiation in various occupational settings.

REFERENCES

Ahmad, M.I., Ab Rahim, M.H., Nordin, R., Mohamed, F., Abu-Samah, A. and Abdullah, N.F., 2021. Ionizing radiation monitoring technology at the verge of internet of things. *Sensors*, 21(22), p.7629.

Albander, H., 2021. Occupational health and radiation safety of radiography workers. In Yongxia Zhou (ed), *Biomedical signal and image processing* (p. 122). United Kingdom: IntechOpen.

Bais, A.F., Lucas, R.M., Bornman, J.F., Williamson, C.E., Sulzberger, B., Austin, A.T., Wilson, S.R., Andrady, A.L., Bernhard, G., McKenzie, R.L. and Aucamp, P.J., 2018. Environmental effects of ozone depletion, UV radiation and interactions with climate change: UNEP Environmental Effects Assessment Panel, update 2017. *Photochemical & Photobiological Sciences*, 17(2), pp.127–179.

Chaturvedi, A. and Jain, V., 2019. Effect of ionizing radiation on human health. *International Journal of Plant and Environment*, 5(03), pp.200–205.

Code, S., 2024. *Safety procedures for the installation, use and control of x-ray equipment in large medical radiological facilities* (Ottawa). ISBN: 978-0-660-72790-5

DeBoom, M.J., 2020. Toward a more sustainable energy transition: lessons from Chinese investments in Namibian uranium. *Environment: Science and Policy for Sustainable Development*, 62(1), pp.4–14.

Dobbinson, S.J., Jamsen, K., Dixon, H.G., Spittal, M.J., Lagerlund, M., Lipscomb, J.E., Herd, N.L., Wakefield, M.A. and Hill, D.J., 2014. Assessing population-wide behaviour change: concordance of 10-year trends in self-reported and observed sun protection. *International Journal of Public Health*, 59, pp.157–166.

D'Orazio, J., Jarrett, S., Amaro-Ortiz, A. and Scott, T., 2013. UV radiation and the skin. *International Journal of Molecular Sciences*, 14(6), pp.12222–12248.

Dube, T.T., 2015. Background radiation in radiography departments and who wears Thermoluminescent Dosimeter (TLD) Badges (Doctoral dissertation).

Dunn, A.M., Burns, C. and Sattler, B., 2003. Environmental health of children. *Journal of Pediatric Health Care*, 17(5), pp.223–231.

Gbetchedji, A.A., Houndetoungan, G.D., Hounsossou, H.C., Journy, N., Haddy, N., Rubino, C., Biaou, O., Medenou, D., Amoussou-Guenou, K.M., de Vathaire, F. and Allodji, R.S., 2020. A systematic review of occupational radiation individual dose monitoring among healthcare workers exposed in Africa. *Journal of Radiological Protection*, 40(4), p.R141.

Golden, A.P., Milder, C.M., Ellis, E.D., Anderson, J.L., Boice Jr, J.D., Bertke, S.J. and Zablotska, L.B., 2021. Cohort profile: four early uranium processing facilities in the US and Canada. *International Journal of Radiation Biology*, 97(6), pp.833–847.

Grison, S., Favé, G., Maillot, M., Manens, L., Delissen, O., Blanchardon, E., Dublineau, I., Aigueperse, J., Bohand, S., Martin, J.C. and Souidi, M., 2016. Metabolomics reveals dose effects of low-dose chronic exposure to uranium in rats: identification of candidate biomarkers in urine samples. *Metabolomics*, 12, pp.1–16.

Guéguen, Y., Roy, L., Hornhardt, S., Badie, C., Hall, J., Baatout, S., Pernot, E., Tomasek, L., Laurent, O., Ebrahimian, T. and Ibanez, C., 2017. Biomarkers for uranium risk assessment for the development of the CURE (Concerted Uranium Research in Europe) molecular epidemiological protocol. *Radiation Research*, 187(1), pp.107–127.

Hadjee, A.S., 2020. Exposure of car guards to solar ultraviolet radiation during summer and winter (Dissertation, North-West University (South-Africa)).

Hanke, R., Fuchs, T. and Uhlmann, N., 2008. X-ray based methods for non-destructive testing and material characterization. *Nuclear Instruments and Methods in Physics Research Section A: Accelerators, Spectrometers, Detectors and Associated Equipment*, 591(1), pp.14–18.

Hecht, G., 2009. Africa and the nuclear world: labor, occupational health, and the transnational production of uranium. *Comparative Studies in Society and History*, 51(4), pp.896–926.

Huang, X. and Chalmers, A.N., 2021. Review of wearable and portable sensors for monitoring personal solar UV exposure. *Annals of Biomedical Engineering*, 49(3), pp.964–978.

Hussain, S., Mubeen, I., Ullah, N., Shah, S.S.U.D., Khan, B.A., Zahoor, M., Ullah, R., Khan, F.A. and Sultan, M.A., 2022. Modern diagnostic imaging technique applications and risk factors in the medical field: a review. *BioMed Research International*, 2022(1), p.5164970.

Jiménez, P., Medlen, K.P. and Fleitas-Estévez, I., 2014. Diagnostic imaging for global health: implementation and optimization of radiology in the developing world. In D.J. Mollura and M.P. Lungren (eds), *Radiology in global health: strategies, implementation, and applications* (pp. 127–137). New York: Springer New York.

Johnston, J., Comello, R.J., Vealé, B.L. and Killion, J., 2010. Radiation exposure dose trends and radiation dose reduction strategies in medical imaging. *Journal of Medical Imaging and Radiation Sciences*, 41(3), pp.137–144.

Julian, A.K., 2017. *Measuring health beliefs about UV exposure among Oregon College students* (Doctoral thesis, Oregon State University).

Kholopo, M. and Rathebe, P.C., 2024. Radon exposure assessment in occupational and environmental settings: an overview of instruments and methods. *Sensors*, 24(10), p.2966.

Krūmiņš, J. and Kļaviņš, M., 2023. Investigating the potential of nuclear energy in achieving a carbon-free energy future. *Energies*, 16(9), p.3612.

Lewis, S., Downing, C. and Hayre, C.M., 2023. Radiation protection among South African diagnostic radiographers—a mixed method study. *Health Physics*, 124(3), pp.208–216.

Lin, X., Luo, J., Liao, M., Su, Y., Lv, M., Li, Q., Xiao, S. and Xiang, J., 2022. Wearable sensor-based monitoring of environmental exposures and the associated health effects: a review. *Biosensors*, 12(12), p.1131.

Linde, K., Wright, C.Y. and Du Plessis, J.L., 2022. Personal solar ultraviolet radiation exposure of farmworkers: seasonal and anatomical differences suggest prevention measures are required. *Annals of Work Exposures and Health*, 66(1), pp.41–51.

Linhares, D., Garcia, P. and Rodrigues, A., 2017. Radon exposure and human health: what happens in volcanic environments? In F. Adrovic (ed), *Radon* (p. 156). London, UK: IntechOpen.
Lucas, R., Prüss-Üstün, A., Perkins van Deventer, E. and World Health Organization., 2010. *Solar ultraviolet radiation: assessing the environmental burden of disease at national and local levels*. Geneva: World Health Organization.
Makumbi, T., Breustedt, B. and Raskob, W., 2024. Parameter uncertainty analysis of the equivalent lung dose coefficient for the intake of radon in mines: a review. *Journal of Environmental Radioactivity*, 276, p.107446.
McCormack, V.A. and Schüz, J., 2012. Africa's growing cancer burden: environmental and occupational contributions. *Cancer Epidemiology*, 36(1), pp.1–7.
Meghzifene, A., Dance, D.R., McLean, D. and Kramer, H.M., 2010. Dosimetry in diagnostic radiology. *European Journal of Radiology*, 76(1), pp.11–14.
Miller, K.A., In, G.K., Jiang, S.Y., Ahadiat, O., Higgins, S., Wysong, A. and Cockburn, M.G., 2017. Skin cancer prevention among Hispanics: a review of the literature. *Current Dermatology Reports*, 6, pp.186–195.
Moshupya, P.M., Abiye, T., Mouri, H. and Levin, M., 2019. The Uranium and Radon gas concentrations and impact on human health: a case from Abandoned Gold Mine Tailings in the West Rand Area, Krugersdorp, South Africa (Dissertation, University of the Witwatersrand, Faculty of Science).
Mphaga, K.V., Mbonane, T.P., Utembe, W. and Rathebe, P.C., 2024. Short-term vs. long-term: a critical review of indoor radon measurement techniques. *Sensors*, 24(14), p.4575.
Mphaga, K.V., Utembe, W., Shezi, B., Mbonane, T.P. and Rathebe, P.C., 2024. Unintended consequences of urban expansion and gold mining: elevated indoor radon levels in Gauteng communities' neighboring gold mine tailings. *Atmosphere*, 15(8), p.881.
Mudd, G.M., 2008. Radon sources and impacts: a review of mining and non-mining issues. *Reviews in Environmental Science and Bio/Technology*, 7, pp.325–353.
Neale, R.E., Lucas, R.M., Byrne, S.N., Hollestein, L., Rhodes, L.E., Yazar, S., Young, A.R., Berwick, M., Ireland, R.A. and Olsen, C.M., 2023. The effects of exposure to solar radiation on human health. *Photochemical & Photobiological Sciences*, 22(5), pp.1011–1047.
Oakley, P.A., Ehsani, N.N. and Harrison, D.E., 2019. The scoliosis quandary: are radiation exposures from repeated X-rays harmful? *Dose-Response*, 17(2), p.1559325819852810.
Olawole, T., Oyetunde, T., Uzomah, U., Shanahan, J., Hartmann, K., Rotimi, S. and Dako, F., 2024. Exploring the state of cancer imaging research in Africa. *Journal of the American College of Radiology*, 21(8), pp.1216–1221.
Othman, S.A., Rosli, N.N.F. and Farizah, N.H., 2023. The effectiveness of radiation protection in medical field-a short review. *Malaysian Journal of Applied Sciences*, 8(1), pp.65–73.
Papadopoulou, P., Misseyanni, A. and Marouli, C., 2020. Current environmental health challenges: Part I-exposures and research trends. In G. Duca and A. Vaseashta (eds), *Handbook of research on emerging developments and environmental impacts of ecological chemistry* (pp. 1–37). New York: IGI Global.
Parr, C.L., 2008. Self-reported exposure data in epidemiologic studies (Doctoral dissertation, University of Oslo Norway).
Passeron, T., Lim, H.W., Goh, C.L., Kang, H.Y., Ly, F., Morita, A., Ocampo-Candiani, J., Puig, S., Schalka, S., Liu, W. and Demessant-Flavigny, A.L., 2023. Sun exposure behaviours as a compromise to paradoxical injunctions: insight from a worldwide survey. *Journal of the European Academy of Dermatology and Venereology*, 37(12), pp.2481–2489.
Raj, K., Prasad, K.K. and Bansal, N.K., 2006. Radioactive waste management practices in India. *Nuclear Engineering and Design*, 236(7–8), pp.914–930.
Rathod, A.M., Verpaele, S., Kelvin, M. and Leybourne, M.I., 2023. Uranium: an overview of physicochemical properties, exposure assessment methodologies, and health effects of environmental and occupational exposure. *Environmental Geochemistry and Health*, 45(5), pp.1183–1200.

Robertson, A., Allen, J., Laney, R. and Curnow, A., 2013. The cellular and molecular carcinogenic effects of radon exposure: a review. *International Journal of Molecular Sciences*, 14(7), pp.14024–14063.

Roro, A.G., Terfa, M.T., Solhaug, K.A., Tsegaye, A., Olsen, J.E. and Torre, S., 2016. The impact of UV radiation at high altitudes close to the equator on morphology and productivity of pea (Pisum sativum) in different seasons. *South African Journal of Botany*, 106, pp.119–128.

Sarman, I., Hassan, D.H. and Sarman, I., 2016. Factors affecting radiographers' compliance with radiation protection on all areas of hospital settings worldwide-a meta-analysis. *International Journal for Innovative Research in Science & Technology*, 3(4), pp.433–438.

Schoonhoven, M., 2012. Occupational exposure to radon in a South African platinum mine (Dissertation, North-West University).

Semenova, Y., Pivina, L., Zhunussov, Y., Zhanaspayev, M., Chirumbolo, S., Muzdubayeva, Z. and Bjørklund, G., 2020. Radiation-related health hazards to uranium miners. *Environmental Science and Pollution Research*, 27, pp.34808–34822.

Sharma, B.R., Singh, S., Mandal, S. and Poudel, N., 2019. Knowledge of radiation exposure and it risk among radiographers and radio technologists. *Journal of Gandaki Medical College-Nepal*, 12(2), pp.17–22.

Signé, L., 2021. *Strategies for effective health care for Africa in the fourth industrial revolution*. Washington DC: Brookings.

Suleiman, A.B. and Muhammad, F., 2022. The modern information technology (IT) in healthcare: an African perspective. *International Journal of Health and Life Sciences*, 8(1), pp.1–4.

Tamakuma, Y., Kranrod, C., Suzuki, T., Watanabe, Y., Ploykrathok, T., Negami, R., Nugraha, E.D., Iwaoka, K., Janik, M., Hosoda, M. and Tokonami, S., 2020. Passive-type radon monitor constructed using a small container for personal dosimetry. *International Journal of Environmental Research and Public Health*, 17(16), p.5660.

Thomas, P., Swaminathan, A. and Lucas, R.M., 2012. Climate change and health with an emphasis on interactions with ultraviolet radiation: a review. *Global Change Biology*, 18(8), pp.2392–2405.

Tootell, A., Szczepura, K. and Hogg, P., 2014. An overview of measuring and modelling dose and risk from ionising radiation for medical exposures. *Radiography*, 20(4), pp.323–332.

Turner, J., Igoe, D., Parisi, A.V., McGonigle, A.J., Amar, A. and Wainwright, L., 2020. A review on the ability of smartphones to detect ultraviolet (UV) radiation and their potential to be used in UV research and for public education purposes. *Science of the Total Environment*, 706, p.135873.

Van Schalkwyk, A., 2005. A radiation monitoring program in a South African gold mine (Doctoral dissertation, North-West University).

Vance, R., Hinton, N., Huffman, D., Harris, F., Arnold, N., Ruokonen, E., Jakubick, A., Tyulyubayev, Z., Till, W.V., Woods, P. and Hall, S., 2014. Managing environmental and health impacts of uranium mining (No. NEA--7062). Organisation for Economic Co-Operation and Development.

Varshney, R., 2013. *Natural radioactivity and Radon/Thoron measurement in environment & quantification of heavy elements*. India: Aligarh Muslim University.

Voldřich, A., 2019. Radon, the silent killer: the neglected repercussions cause unforeseen environmental health risks in African countries (Dissertation, Wien).

Wallin, A., Gustafsson, M., Anderzen Carlsson, A. and Lundén, M., 2019. Radiographers' experience of risks for patient safety incidents in the radiology department. *Journal of Clinical Nursing*, 28(7–8), pp.1125–1134.

Yi, L., Gao, B., Liu, H., Zhang, Y., Du, C. and Li, Y., 2020. Characteristics and assessment of toxic metal contamination in surface water and sediments near a uranium mining area. *International Journal of Environmental Research and Public Health*, 17(2), p.548.

2 Ionizing radiation exposure from mine tailings

2.1 BACKGROUND OF THE CHAPTER

Gold mining has been a significant economic activity in Africa for centuries, contributing to the continent's wealth and development (Campbell 2009). However, the environmental and health impacts of gold mining, particularly in relation to uranium exposure from mine tailings, have become increasingly concerning (Srivastava et al. 2020). Uranium, a naturally occurring radioactive element, is often found in association with gold deposits and can pose serious health risks to humans and the environment. Exposure to uranium from gold mine tailings can occur through various pathways, including inhalation of dust containing uranium particles, ingestion of contaminated food and water, and direct contact with contaminated soil and water (Zupunski et al. 2023). In the body, uranium can accumulate in organs such as the kidneys and bones, leading to long-term health effects such as cancer and kidney damage (Brugge et al. 2005).

Africa is home to some of the largest gold mining operations in the world, with countries like South Africa, Ghana, and Mali being major producers. These mining activities have the potential to release significant amounts of uranium into the environment, particularly through the disposal of mine tailings. As a result, communities living near gold mines are at risk of exposure to elevated levels of uranium. Despite the potential health risks associated with uranium exposure from gold mine tailings, there is a lack of comprehensive studies on this subject in Africa. Existing research has mainly focused on the environmental impact of gold mining, with limited attention paid to uranium exposure and its health effects on nearby communities. While mine tailings release uranium, radon is also one of the naturally occurring radioactive gases that is produced by the decay of uranium and thorium in the Earth's crust (Khattak et al. 2011). It is a colourless, odourless gas that can accumulate in indoor spaces, particularly in areas with poor ventilation (Mphaga et al. 2024). Radon exposure is a significant environmental and public health concern. It remains the second leading cause of lung cancer worldwide, responsible for an estimated 15% of lung cancer deaths annually (Riudavets et al. 2022). In Africa, radon exposure is of particular concern in regions with extensive mining activities, such as gold mining areas. Mining operations can release radon gas into the air and water, leading to elevated radon levels in residential areas near mines (Mphaga et al. 2024). Additionally, certain geological formations in Africa, such as granite rocks, can also contribute to higher radon levels in indoor spaces (Le Roux et al. 2019).

DOI: 10.1201/9781003508588-2

Despite the known health risks associated with radon exposure, there is limited awareness and research on this issue in many African countries. Most studies have focused on high-income countries, with little attention paid to radon exposure in Africa. This is concerning, as Africa is home to a large and growing population, many of whom may be at risk of radon exposure. In this chapter, sources and pathways of uranium and radon exposure, the health exposure-effects, and regulatory framework governing uranium and radon exposure are discussed, highlighting the environmental and health impacts from gold mine tailings.

2.2 OVERVIEW OF GOLD MINE TAILINGS IN AFRICA

Gold mine tailings are the residual materials left over after the extraction of gold from ore. These tailings consist of finely ground rock particles and chemicals used in the gold extraction process, such as cyanide and mercury (Johnson 2015). They are typically stored in large impoundments or tailings dams near the mining site. From an environmental health perspective, gold mine tailings pose several significant risks to both human health and the environment. One of the primary concerns is the potential for these tailings to leach toxic chemicals and heavy metals into the surrounding environment (Zhang et al. 2021). This usually occurs through the erosion of the tailings dam, seepage into groundwater, or runoff into nearby water bodies. Leaching can result in the contamination of surface water and groundwater with toxic chemicals and heavy metals (Abdul-Wahab and Marikar 2012). This can lead to water sources being polluted and unsafe for human consumption and for use in agriculture, posing a significant risk to human health and crop production, through irrigation. This may present potential risks for health problems such as cancer, respiratory diseases, and neurological disorders (Ngole-Jeme and Fantke 2017). The effects of gold mine tailing leach can persist for many years, even after mining activities have ceased, resulting in ongoing environmental and health risks for local communities.

The inhalation of dust containing heavy metals and other contaminants can lead to severe respiratory health challenges (Rehman et al. 2018), while consumption of water and food products contaminated with heavy metals from mine tailings can result in long-term health problems, including cancer and neurological disorders (Sonone et al. 2020). The impact of gold mine tailings on the environment is also profound. The physical presence of these tailings can disrupt ecosystems, leading to soil erosion, loss of biodiversity, and habitat destruction (Timsina et al. 2022). Additionally, the release of chemicals and heavy metals into water bodies can harm aquatic life and disrupt entire ecosystems (Shanmukha et al. 2024). Efforts to mitigate the environmental and health risks associated with gold mine tailings in Africa are ongoing (Sibanda 2019). These efforts include the implementation of stricter regulations on tailings management, the development of safer mining practices, and the remediation of contaminated sites. However, addressing these challenges requires a multidisciplinary approach that involves government agencies, mining companies, and local communities working together to protect both the environment and human health.

FIGURE 2.1 Gold mine tailing

2.3 PROXIMITY OF GOLD MINE TAILINGS TO RESIDENTIAL STRUCTURES

Until the 1950s, gold mining waste in Africa was typically deposited in sparsely populated areas characterized by savannah grassland known as veldt, featuring grass, shrubs, and few trees (Figure 2.1). At the time, residential areas were located approximately 3–4 km away from the edges of the gold mine tailings (Figure 2.2). However, the rapid expansion of mining from the 1950s onwards resulted in the socio-economic growth of mining areas in Africa. This led to swift urbanization to the mining towns and the simultaneous establishment of dormitory suburbs in some areas of Africa, particularly Southern Africa, due to the forced resettlement policies of the colonization and apartheid-era regimes.

Post-apartheid and colonization, there was a significant increase in rural-urban migration in Africa as people sought better livelihoods. This migration led to increased development of formal and informal housing (Oberhauser 2016). Currently, there is a housing backlog in South Africa and other African countries, and there is increasing pressure for more housing developments, particularly on remaining open land within the greater city limits along the Witwatersrand mining corridor in South Africa. These historical developments have shaped housing patterns and dynamics in Africa, resulting in extensive infill of vacant land that once acted as a safety buffer zone near both active and dormant mine tailings. This has increased exposure to health hazards and environmental consequences of dust pollution from the gold mine tailings.

Mine tailings, often referred to as mine waste dumps or colloquially as mine dumps, are sites where mine tailings are stored. Tailings are a byproduct of the process of separating valuable minerals from uneconomic material and consist of ground-up rock, process water, and chemical reagents (Owen et al. 2020). These facilities can be classified into three groups based on their genesis: hydraulic-filled dumps, piled dumps, and technogenic dumps (Pashkevich 2017). Hydraulic-filled dumps arise from aggradation materials from engineering activities, while technogenic

FIGURE 2.2 Proximity of gold mine tailings to residential activities

dumps result from industrial activities and are characterized by low thickness and considerable area. Piled dumps are formed from stacked materials, including soils of natural origin, mineral waste, or solid industrial waste. Mining activities generate various types of waste, such as hard rocks, gravels, clays, pebbles, sands, limestones, and flotation tailings of ferrous and nonferrous metal ores, among others (Pashkevich 2017).

The volume of mining waste produced depends on factors such as ore-processing technology, geographical location of mines, and types of minerals extracted (Anawar 2015). For instance, smelting 1 tonne of copper produces 4.2 tons of tailings and 30 tons of ash, while obtaining 1 ton of gold involves processing 23 million tons of rock mass (Gümüşsoy et al. 2023). Most mine waste dumps are unlined and unvegetated, making them significant sources of silica-rich airborne dust, which consists of fine particulate matter (Maseki et al. 2017). Reworking of gold mine dumps for further gold extraction using newer technologies increases the proportion of inhalable particles in the tailing material. Mine dusts contain minerals such as quartz, carbonates, sulphates, phosphates, and feldspars, with their composition indicating their source region. In Witwatersrand gold mines, silica makes up 75–85% of the mine dumps, followed by pyrite at 1.5–2.5%. Ores of copper, lead, zinc, coal, gold, and uranium often contain hazardous mineral elements, with sulphides being the most common. Toxic elements such as arsenic, uranium, lead, zinc, cadmium, mercury, and others are also present, with higher proportions in the dumps compared to the ore. Dust pollution on mine dumps occurs during the deposition of fine dry materials and heavy winds if the dump surfaces are dry. The fine particles in sand and slime dams lack binding or water retention properties, making them highly susceptible to wind erosion and generating dust or air pollution.

In South Africa, particularly the Witwatersrand in Johannesburg, there are approximately 270 mine tailings in the region that brought several issues, such as a continued surge in migrant labourers, great land pressure, and public health concerns related to exposure to airborne mineral quartz dust (Kneen 2014). Although no study has attempted to quantify mine tailings in the entire African region, all challenges that are brought by these tailings have become the focus of research initiatives. A study by Maseki (2013) selected three large mine tailings in Southern Africa, Johannesburg, to investigate historical changes in population density within the zones of maximum dust exposure: the Crown Gold Recoveries (CGR) complex in central Johannesburg, the Durban Roodepoort Deep (DRD) tailings storage facility (TSF) on the west Witwatersrand, and the East Rand Proprietary Mine (ERPM) TSF complex on the east Witwatersrand. These TSFs, each with an aerial footprint exceeding 1 km^2, are the largest three mine tailings in the Central Witwatersrand goldfield and represent a range of conditions.

Mining activities in Africa create a demand for housing due to the need for labour. Historically, mining companies were responsible for providing employee housing. However, in Africa, this practice changed in the mid-1980s and 1990s for several reasons. Declining resource prices led to a shift in focus away from peripheral activities like housing, concerns arose that mining companies were assuming a role that should be handled by the government, and new labour regulations limited investment in housing. Africa's mining history is intertwined with its colonial and post-colonial development. The colonial legacy often included the establishment of mining towns to house workers, which continued in the post-colonial era. Post-independence, there was pressure on mining companies to consider different housing ownership models, leading to the upgrade of compounds to single-living quarters. However, this policy resulted in fewer employees being accommodated and contributed to the development of informal settlements around mines.

Apartheid-like policies also influenced the establishment of communities near mines in some African countries. For example, in South Africa, the apartheid regime controlled where different racial groups could live, leading to the creation of townships near mines for black and coloured workers. These townships often lacked basic infrastructure and services, leading to poor living conditions. Today, many communities in Africa live in close proximity to mine waste facilities, which contain reactive minerals and toxic metals, posing significant risks to their health and well-being. It is crucial for governments and mining companies to address the housing needs of mining communities and mitigate the environmental and health risks associated with living near mines.

2.4 OVERVIEW OF LITERATURE ON RADON AND URANIUM RESIDENTIAL EXPOSURE PROXIMAL TO GOLD MINE TAILINGS

2.4.1 Radon

Several studies have investigated radon levels in areas proximal to gold mine tailings to assess the potential health risks to nearby residents. For example, a study conducted by Moshupya et al. (2023) in Gauteng measured radon levels in homes located near gold mine tailings and compared them to levels in homes located further away. The study found that homes near the tailings had significantly higher radon levels, indicating that the tailings were a potential source of radon exposure for residents in the area. Another study by Mphaga et al. (2024) conducted a similar investigation in a different region near gold mine tailings. This study also found elevated radon levels in homes near the tailings, with levels exceeding the recommended guidelines set by the World Health Organization (WHO) for indoor radon concentrations (IRCs). These findings suggest that residents living near gold mine tailings may be at increased risk of radon exposure and associated health effects.

Exposure to radon poses significant health risks to humans. Radon undergoes decay to form 218Po and 214Po, which are solid particles that attach to aerosol and dust particles in the air. When these particles are inhaled, they irradiate the lung lining tissues with alpha particles, leading to lung cancer (Degu Belete and Alemu Anteneh 2021). Radon is currently recognized as the second leading cause of lung cancer (Riudavets et al. 2022). The likelihood of developing lung cancer due to radon exposure is influenced by factors such as the dose received, duration of exposure, smoking habits, and exposure to other lung cancer-causing agents (Liu et al. 2024). Individuals exposed to high radon levels face a higher risk of lung cancer, while long-term exposure to low concentrations also increases the risk (Schmid et al. 2010).

Indoor radon levels are influenced by various sources and factors. Radon is primarily derived from the decay of uranium (238U) and can come from sources such as uranium- and radium-bearing soils and rocks, building materials, domestic water, natural gas, and outdoor air. Higher uranium concentrations in an area increase the likelihood of high radon levels indoors (Sundal et al. 2004). IRCs are often elevated in areas with granites, shales, soils, and sediments rich in uranium and radium. Human activities such as mining and processing of uranium, gold, phosphorus, and coal combustion can also increase radon levels in indoor environments. Once radon enters indoor spaces, it becomes trapped and accumulates, with the amount accumulating dependent on the building's size and ventilation rate (Bulut and Şahin 2024).

In Africa, particularly South Africa historic gold mining activities have led to the deposition of large tailings dams enriched with radioactive elements, which are potential sources of radon (Mphaga et al. 2024). Studies in the West Rand region, particularly Krugersdorp, have shown outdoor radon levels ranging from 32 Bq/m^3 to 1,069 Bq/m^3 near gold mine tailings (Moshupya et al. 2019; Moshupya et al. 2023). Some mine tailings in the region have high uranium-238 activity concentrations, the parent radionuclide of radon, raising concerns about radon dispersion into nearby dwellings (Mphaga et al. 2024). The proximity of many dwellings to these tailings dams in the gold mine regions of South Africa increases the risk of elevated IRCs from radon emanating from the tailings and erosion of tailings material into residential areas (Komati et al. 2022).

International agencies responsible for setting guidelines on radiological protection and policies to safeguard public health have defined reference levels for indoor radon exposure. The International Atomic Energy Agency (IAEA) (2019) and the International Commission on Radiological Protection (ICRP) (1994) recommend a reference level of 300 Bq/m^3 for indoor radon exposure. In contrast, the WHO suggests a lower reference level of 100 Bq/m^3 to reduce health risks associated with indoor radon exposure and specifies that if this level cannot be met, the indoor radon value should not exceed 300 Bq/m^3 (2009). In one of the studies conducted in South Africa, an analysis of indoor radon measurements conducted during the summer months showed that the average activity concentrations in all measured dwellings were below the recommended reference levels of 300 Bq/m^3 and 100 Bq/m^3 proposed by the ICRP, IAEA, and WHO. During winter season, the indoor radon levels were also below the recommended reference level of 300 Bq/m^3 (Moshupya et al. 2023).

In Johannesburg, South Africa, a study conducted by Moshupya et al. (2019), took samples from rocks, tailings, and construction materials for geochemical analyses. Radon monitors (60) were installed in indoor and outdoor environments for radon characterization, and the findings revealed elevated uranium levels in mine tailings, with a maximum of 149.76 ppm and a mean value of 48.87 ppm. Radon levels in the area ranged from 32 Bq/m^3 to 1069 Bq/m^3, exceeding the typical outdoor radon levels of about 10 Bq/m^3, with particularly high values observed in gold tailings dams. Table 2.1 provides an overview of radon exposure studies conducted in South Africa between 1989 and 2019. IRCs reached a maximum of 174 Bq/m^3, surpassing the recommended value of 100 Bq/m^3. Calculations of effective doses based on these radon levels indicated a maximum of 16 mSv/y, well above the recommended public exposure limit of 1 mSv/y. Samples of water, soil, ore, mine tailings, and air were analysed using gamma spectrometry, while radon concentrations were measured using solid-state nuclear track detectors in a Ghanaian study (Darko et al. 2010). A survey was conducted to determine the ambient gamma dose rate in the vicinity of the mines and surrounding areas. Effective doses resulting from external gamma irradiation, ingestion of water, and inhalation of radon and ore dusts were calculated for the two mines. The average annual effective dose was determined to be 0.30 ± 0.06 mSv, a result consistent with levels reported in other countries. Table 2.1 provides an overview of studies conducted between 1989 and 2019 in South Africa.

Research on radon exposure in Africa is relatively limited compared to other regions. However, some studies have been conducted, particularly in regions where radon exposure is a concern due to geological factors. The study on radon measurements in dwellings of northern Algeria aimed to assess radon concentrations in residential buildings using passive detectors (Cherouati and Djeffal 1988). The results revealed wide variations in radon levels, with some areas showing concentrations exceeding recommended guidelines. Factors such as building materials, ventilation, and geological characteristics were found to influence radon levels. The study highlighted the importance of monitoring radon levels in residential buildings to assess potential health risks. Mitigation strategies, such as improving ventilation and using radon-resistant construction materials, were suggested to reduce radon exposure. Overall, the study emphasized the need for further research and awareness campaigns to address radon exposure in northern Algeria and similar regions.

Researchers (Obed et al. 2012) conducted indoor radon measurements across various regions of South-Western Nigeria using CR-39 nuclear track detectors. The study assessed radon levels in different types of dwellings, providing insights into the spatial distribution of radon concentrations. Findings highlighted potential health risks associated with radon exposure in residential buildings, emphasizing the need for further research and mitigation efforts. In a study, Mbida Mbembe et al. (2023) focused on indoor radon measurements in Cameroon, the Ebolowa town, using passive detectors. The study identified factors such as building materials and ventilation systems as the main contributors of high radon levels in residential buildings. Setso (2020) investigated indoor radon levels in selected areas of Serule, Botswana, using solid-state nuclear track detectors. The results of the study showed varying

TABLE 2.1
Indoor radon exposure assessment studies between 1989 and 2019 in South Africa

Location	Reference	Year	Methods	Measurement period	Average concentration (Bq/m³) found
Witwatersrand Cape town Pretoria	Leuchner, van As, Grundling and Steyn (1989)	1989	Track-etch samplers and charcoal canisters	3 months Passive measurements	40 20 300 600
Paarl region near the Berg River (west) (East)	Lindsay et al. (2008)	2008	Electret ion Chambers (E-PERM) (Passive technique)	5–20 days in the living room Living room	132 37
Midvaal Water, Klerksdorp Botshabelo, Klerksdorp	Kgabi et al. (2009).	2009	Radon Electret Chambers (S chamber, E-PERM type). Passive technique	25 days 21 days	30.16 ±2.52 pCi/L (1,115.92 Bq/m³) 46.06 ± 5.21 pCi/L (1,704.220 Bq/m³)
Ezulweni mine West Rand	Dehkordi (2012)	2011	Alpha Guard active detector RGMs Passive (SSNTND) technique	7 days 2 months winter at night and early morning	Winter 44.67 Summer 33.17 Winter 38.58 Summer 30.83 Average 38
Carletonville	Kamunda et al. (2017)	2017	Alpha Guard Professional Radon Monitor (active technique)	24 hours over 3 days	119.5
Krugersdorp West Rand region	Moshupya et al. (2019)	2017–2018	Solid-state nuclear track device (RGMs)	3 months	105
Baviaansberg	Bezuidenhout (2019)	2018	Sodium iodide (NaI (Tl)) scintillation	5 minutes Summer	400
Vredenburg Saldanha	Le Roux et al. (2019).	2019	Electret ion chamber (Passive technique)	3 days	58.7 38.6

Adopted from Ocwelwang et al. (2019).

radon concentrations between dwellings, as well as the effectiveness of ventilation systems in reducing radon accumulation. In central Cameroon, a comparative analysis of indoor radon measurements and dose assessments between residential buildings and inhabitants revealed high potential health risks associated with radon exposure (Ndjana Nkoulou et al. 2022). While direct studies on radon exposure, exposure assessments, and pulmonary health risks are limited, exposure and health studies are urgently needed to investigate the relationship between radon exposure and chronic respiratory effects including lung cancer in specific regions of Africa, particularly in areas where mining activities are prevalent.

Radon-222 (222Rn) is a naturally occurring, inert radioactive gas produced directly from the alpha decay of radium-226 (226Ra), which is part of the uranium-238 (238U) decay chain. This gas is colourless and odourless, undetectable by human senses, and has a half-life of 3.82 days (Setso 2020). Historically, it was believed that significant radon exposure was confined to underground mines and their workers (Sahu et al. 2016). However, numerous international studies have shown that residents and workers in above-ground environments near gold or uranium mines and regions with high granite content can also encounter elevated radon levels that surpass recommended limits (Voldřich 2019).

Radon is a significant source of radiation exposure for the public, including in homes globally (Mphaga et al. 2024). Inhalation or ingestion of radon progenies can damage DNA and induce lung cancer (Robertson et al. 2013). The International Agency for Research on Cancer (IARC) has confirmed that prolonged exposure to radon and its decay products, such as polonium-218 (218Po) and polonium-214 (214Po), leads to lung cancer in humans. Health impact studies on radon in underground workplaces began decades ago, primarily focusing on uranium miners (Tsewu 2021). By the late 1970s and early 1980s, these studies expanded to include radon concentration measurements in private homes and other high-occupancy buildings. Therefore, indoor radon measurements are crucial for assessing human exposure in homes and public buildings.

South Africa has a long history of gold mining, with gold and uranium extracted from the Witwatersrand Basin since 1886. This basin includes several regions like Central Rand, Eastern Basin, Western Basin, Far West Basin, Free State gold mines, and the Klerksdorp goldfields (KOSH region). A report by Liefferink (2019) highlighted that these mines have created over 270 TSFs containing about 600,000 tons of uranium. Consequently, gold mine tailings in South Africa are expected to exhibit high radon levels, given that uranium, a long-term radon source, is a byproduct of gold mining (Wymer and Botha 2001). Additionally, these TSFs contain other toxic heavy metals associated with gold mining. Despite numerous studies on the impacts of gold mining, there is a notable lack of indoor radon research in the country. Thus, there is a critical need for indoor radon surveys and continuous monitoring in residential areas near gold and uranium mine tailings to assess the health risks to nearby communities.

The significant health risks associated with high indoor radon levels are well-documented in Europe and the United States but with less quantification in Africa. However, measuring radon levels in homes is expensive and time-consuming. Main parameters influencing indoor radon levels include the underlying geology and climatic conditions, particularly temperature and wind. Different geological formations result in varying radon exhalation rates, while climate can either mitigate or exacerbate indoor radon levels (Abed et al. 2024). Mbida Mbembe et al. (2023) presented measurements of indoor Radon-222 activity concentration, external gamma dose rates, and cancer risk assessment in various dwellings of Ebolowa town, Southern Cameroon. The region's geology is primarily composed of granites, and it lies within the equatorial climatic zone. The area's mean temperature ranges from 25°C to 26°C, experiencing two rainy and two dry seasons with a relative humidity of 70–80% throughout the year. Indoor radon activity concentrations were measured using RADTRACK2 devices, alpha track detectors made of CR-39 plastic films contained in antistatic black holders. Each device was placed in an inhabited room of selected dwellings for approximately three months. Radon concentrations in the homes ranged from 23 ± 10 Bq/m^3 to 2620 ± 480 Bq/m^3, with an arithmetic mean of 135.6 Bq/m^3 and a geometric mean of 63 Bq/m^3. The estimated annual effective doses from external gamma exposure ranged from 0.15 ± 0.01 mSv/y to 0.38 ± 0.06 mSv/y, with an average value of 0.26 mSv/y. The annual effective dose due to indoor radon varied between 0.43 mSv/y and 49.6 mSv/y, with an arithmetic mean of 2.6 mSv/y. The excess lifetime cancer risk (ELCR) ranged from 1.39% to 158%, with a mean value of 8.2%.

In the Eastern Cameroon, Dieu Souffit et al. (2022) investigated the concentrations of radon (222Rn) indoors and in the soil, along with their ambient dose equivalent rates, in the cobalt-nickel bearing region of Lomié. For indoor 222Rn measurement, 98 radon track detectors were used. Soil radon levels were measured with a MARKUS ten detector, and ambient dose equivalent rates were assessed using a RadEye PRD-ER pocket survey meter. Activity concentrations of 226Ra were measured using a high-purity germanium detector (HPGe). The study evaluated annual inhalation and external exposure doses, characteristics of building materials, and correlations between indoor 222Rn, soil 222Rn, and soil 226Ra. IRCs ranged from 30 to 300 Bq/m^3, with an arithmetic mean of 58 Bq/m^3, all exceeding the world average of 30 Bq/m^3 as reported by the United Nations Scientific Committee on the Effects of Atomic Radiation (UNSCEAR). Soil radon concentrations ranged from 3.6 to 63.2 kBq/m^3, with an arithmetic mean of 25.8 kBq/m^3. The annual external effective dose varied from 0.23 to 0.72 mSv, with an arithmetic mean of 0.41 mSv, which is below the global average of 0.9 mSv. A moderate correlation (R2 value of 0.69) was found between 222Rn and 226Ra in the soil. The excess lifetime cancer risk (ELCR) ranged from 1.5% to 16.7%, with an average of 3.6%, which is three times the action level of 1.3% recommended by the United States Environmental Protection Agency (USEPA). In Ghana, a solid-state nuclear track detector (LR-115 type II) was placed in 82 homes within a suburb for three months (September 2017 to January 2018). The IRC ranged from 4.1 to 176.3 Bq/m^3, with an average of 57 ± 39 Bq/m^3. The mean radon exposure for residents was recorded as 0.12 ± 0.08 WLM/y, resulting in an effective lung dose of 0.7 ± 0.5 mSv/y and an excess lifetime cancer risk of 0.39 ± 0.26%. A positive correlation was observed between IRC and building type, with the association being significant (P = 0.047). Salupeto-Dembo et al. (2020) examined radon and thoron activity concentrations in adobe houses in Angola. Passive detectors were used to measure these concentrations during both the rainy and dry seasons in 40 dwellings across three regions with distinct climatic and geological backgrounds: Cabinda, Huambo, and Menongue. Results indicate that radon activity concentrations are higher during the rainy season compared to the dry season, with both radon and thoron levels being the highest in Huambo, located in the central part of the country. The study also estimated the number of adobe houses exceeding certain radon and thoron levels and found that thoron significantly contributes to the inhalation dose.

2.4.2 Uranium

Uranium is naturally found in the Earth's crust, soil, plants, and rocks, with an average abundance of 1.8 mg/kg (Haakonde et al. 2020). It is primarily composed of the 238U isotope (99.27%), with 235U and 234U isotopes making up 0.72% and 0.0055%, respectively (Starościak and Rosiak 2015). Although U occurs naturally as a trace element, it has no known physiological functions in the human body (findings from the literature are inconclusive) and can be radio- and biochemically toxic (Kumar et al. 2015). Mining and smelting are the primary human activities contributing to uranium (U) contamination in water and soil (Giri and Singh 2017). Additional sources include phosphate fertilizers and waste from electronics industries and ammunition factories. Uranium can easily enter aquatic systems through river flow and surface runoff, making its enrichment in groundwater a global concern (Tanwer et al. 2024). Contaminated drinking water, with U levels exceeding 30 μg/L, poses a significant health risk to residents in affected regions (Ma et al. 2020).

Both the WHO and the U.S. Environmental Protection Agency (USEPA) have set a guideline value of 30 μg/L for U in drinking water (WHO 2004), which is higher than Germany's threshold of 10 μg/L (Seldén et al. 2009). It is estimated that over 130 million people worldwide could be exposed to U through drinking water (WHO 2001). Studies across Asia, Europe, and the Americas have shown that groundwater near mines and military dumpsites often contains U levels far exceeding the WHO recommended limit (Singh et al. 2013; Garg et al. 2014; Eggers et al. 2018). Pollution of air, water, and soil from uranium (U) mining poses significant public health risks. Elevated levels of U in groundwater (>30 μg/L) have been documented globally, including in Canada (Alam and Cheng 2014), India, and Mongolia (Kumar et al. 2011; Nriagu et al. 2012). In Africa, U mining and exploration have surged likely due to stricter mining regulations in more developed countries (Besada and Martin 2015). Table 2.2 provides U mining in various African countries.

2.5 FINDINGS FROM AFRICAN STUDIES

In South Africa, gold mine activities have led to the contamination of residential areas with uranium-rich waste from mine tailings. This poses a significant health risk to residents due to uranium's hazardous chemotoxic and radiological properties. Zupunski et al. (2023) conducted a cross-sectional study to assess U concentrations in individual hair samples from children and adults living near mine tailings in Johannesburg, South Africa. The study looked at the association between U concentrations in hair and residential zone, socio-demographic characteristics, and housing conditions. Study sites were categorized into three zones based on the

TABLE 2.2
African countries with Uranium deposits

African countries	Mining production value [% of GDP]	Total export contribution of mining in 2012 [%]
Mauretania	57.5	71.8
Guinea	34.7	87.4
Botswana	29.4	92.4
Zambia	26.4	70.4
Zimbabwe	21.4	38.5
Congo, DR	18	92.4
South Africa	17.1	41.7
Burkina Faso	16	46.3
Namibia	11.6	54.4
Tanzania	8.9	36.7
Gabon	6	85
Mozambique	4.7	47.5
Morocco	3.3	17.2
Central African Republic	3.2	44.5
Niger	1.6	57.5
Madagascar	0.5	20.4
Sudan	0.3	80
Algeria	0.2	98.6
Chad	0	93.8
Equatorial Guinea	0	95.1
Malawi	0	6.2
Average Africa	**12.4**	**60.8**

Adopted from Winde et al. (2017).

distance from dwellings to mine tailings (zone 1: ≤ 500 m, zone 2: 2–3 km, zone 3: 4–5 km). Uranium concentrations in hair samples were measured using inductively coupled plasma-mass spectrometry (ICP-MS). Among 128 participants, 63 (49%) were children aged 7–15 years (38 girls), and 65 (51%) were adult females. The mean (median) U concentration in hair samples was 143 (92) µg/kg. In the adjusted analyses, only an inverse association between age and U concentration in hair was statistically significant, with geometric mean U levels being 2.1 times higher in children compared to adults ($P < 0.001$). Uranium (U) measurements in water, soil, and food linked to gold mining activities in Gauteng Province, South Africa, indicate possible exposure levels that may lead to adverse health effects, including cancer. While theoretical considerations on significant exposure pathways exist, data on direct human exposure are limited. Winde et al. (2019) undertook a cross-sectional study to compare U levels in humans with those from other global settings (based on literature review), explore potential exposure variability within the province, and assess the feasibility of recruiting participants from vulnerable and hard-to-reach populations. Wards with potentially high (HE) and low exposure (LE) were identified. Composite hair samples from regular customers of selected barber shops were collected over

1–2 months, representing the local populations. A total of 70 U concentrations were determined from 27 composite samples comprising 1,332 individuals. U concentrations ranged from 31 μg/kg to 2524 μg/kg, with an arithmetic mean of 192 μg/kg (standard deviation, 310 μg/kg) and a median of 122 μg/kg. Although HE wards collectively showed higher U levels than LE wards (184 vs. 134 μg/kg), the differences were smaller than expected. The study concluded that detected U levels were higher than those reported in most other surveys of the general public. The barber-based approach proved efficient for hair collection. However, due to technical challenges in measuring U, composite hair samples were not recommended.

In Siavonga, Zambia, Haakonde et al. (2021) determined the concentrations of U in the urine samples of Zambians using ICP-MS. The study also correlated urinary U concentrations, U concentrations in drinking water, and cow milk from previous studies conducted in the same study site. The urinary U concentration for residents from communities situated near the U-mining sites, i.e., the U-mining area, had a median concentration of 60.67 μg/L. In comparison, residents from communities located further away from the U-mining sites, i.e., the non-mining area, had urinary U concentrations with a median of 0.72 μg/L. Positive correlations with urinary U concentration were observed for U concentrations in drinking water ($rs = 0.64, p < 0.05$) and cow milk ($rs = 0.63, p < 0.05$), while a negative correlation was noted for age groups ($rs = -0.21, p < 0.05$) of the study participants. The findings indicated elevated exposure levels to U, particularly among the residents in the U-mining area in Siavonga. In Namibia, Agius et al. (2023) investigated the relationship between radiation doses and selected cancers in employees of Rössing Uranium Limited (RUL). The study included individuals with at least one year of continuous employment between 1976 and 2010. Incident cancer cases (lung, extra-thoracic airways (ETA), leukaemia, brain, and kidney) before the end of 2015 were identified from the Namibian and South African National Cancer Registries, as well as RUL's occupational health services. The study included 76 cancer cases (32 lung, 18 ETA, 8 leukaemia, 9 brain, 9 kidney) and a sub-cohort of 1,121 individuals sampled from 7,901 RUL employees. A weighted Cox model, adjusted for available known confounders, produced a rate ratio (95% CI) for lung cancer of 1.42 (0.42, 4.77) and 1.22 (0.26, 5.68) for medium and higher cumulative lung dose categories compared to the lower category, and 1.04 (0.95, 1.13) for a dose increase of 10 mSv. However, the study faced significant challenges with respect to case ascertainment, exposure estimates, and ensuring the accuracy of key variables. Persuasive consistent evidence for elevated cancer risk was not found for radiation or other exposures studied at the Rössing uranium mine.

Njinga et al. (2016) measured the levels of natural uranium contamination in surface water near the Princess Gold Mine in Roodepoort, South Africa were using ICP-MS. The highest activity level of 6.39E+04 mBq/L was found in the reddish brown, ochre surface water from tailings (SWA-RB) close to the houses, while the lowest value of 1.92E+03 mBq/L was found in the flowing surface water (SRWA-5) 1 km away from the dump. Intermediate values of 1.56E+04, 1.07E+04, 1.57E+04, and 8.46E+03 mBq/L were reported at SRWA-2, SRWA-3, and SRWA-4 along the path, respectively. The community living around the tailings uses the surface water for daily consumption. Based on the World Health Organization's recommended annual limit guideline for drinking water (731 L/year), this study found that the community around this area receives a highest annual collective effective dose of 2.10 mSv due to 238U in the drinking surface water. The radiological-health risks associated with 238U in the water samples analysed showed the highest cancer mortality and morbidity values of 2.40E+03 and 3.67E+03, respectively. The mean chemical toxicity risk for natural uranium over a lifetime of consumption was 5.31E+05 µg/kg/day, indicating that the main human risk is likely due to the chemical toxicity of natural uranium. In the Vaal region of Gauteng, rietspruit river flows through residential and agricultural land. Raji et al. (2023) analysed water samples to determine the concentrations of dissolved U. The highest recorded concentration of U was 781.9 µg/L at the mine, with decreasing concentrations observed further from the mine. The maximum annual effective dose at the mine exceeded the regulatory limit for public exposure across all age groups. The radiological risk associated with consuming untreated water from the rietspruit was found to be lower than the chemical risk. Chemical risk levels ranged from 0 to 37.2, with the highest lifetime cancer risk estimated at 1.7 E–03 for morbidity cancer risk and 1.1 E–03 for mortality cancer risk. The chemical risk analysis revealed that within 24 km downstream of the mine, stream users are exposed to toxic levels of uranium.

Höllriegl et al. (2011) assessed the urinary excretion levels of uranium in individuals from Southwest Nigeria residing in areas with low environmental uranium. Participants of various ages and genders were asked to provide 24-hour urine samples, which were analysed for uranium concentration using ICP-MS. Urinary creatinine values were also measured to normalize renal uranium concentrations. Urine uranium concentrations ranged from < 10.4 to 150 ng/L (median 13.8 ng/L), and creatinine-normalized values ranged from 2.52 to 252.7 ng/g creatinine (median 33.4 ng/g creatinine) for adults aged 15 years and older. Children under 15 years exhibited increased uranium excretion, with median values of 61.6 ng/L and 76.0 ng/g creatinine. Daily uranium excretion was estimated at a median of 14.2 ng/d for adults and 45.1 ng/d for children. The uranium excretion levels observed in males and females living in non-mining areas of Nigeria were comparable to reference values reported in other countries with low environmental uranium levels. Stream sediments and groundwater samples were collected near the El Allouga uranium mine in southwestern Sinai, Egypt to investigate their radionuclide content, assessing geochemical dispersion and environmental impact by Ramadan et al. (2022). Radioactivity levels were analysed using γ-ray spectrometry and UV-photometry. Most stream sediment samples showed higher eU concentrations compared to background levels. The presence of significant correlations between eU, clay, and organic matter contents suggests potential U adsorption onto clay and organic matter surfaces. Elevated radionuclide levels in stream sediments are primarily attributed to mining-related contamination, and, in certain areas, to weathering of rock outcrops. The measured U concentrations in groundwater samples exceeded the Maximum Contamination Level of 30 ppb for groundwater U. The lack of correspondence between U concentrations in country rocks and associated groundwater indicates high U mobility and suggests a lack of simple rock/water equilibration. Water resources in the study area displayed 234U/238U activity ratios that deviated noticeably from secular equilibrium, supporting the possibility of local uranium ore body formation within the rock aquifer at El Allouga. The calculated external hazard parameter values exceeded the worldwide average in 30% of the analysed stream sediment samples, indicating an elevated cancer risk for individuals exposed to such radiation levels over a lifetime.

In Nigeria, the concentrations of uranium and thorium were analysed in soils and mineral sands from the Bisichi tin mining area in Jos Plateau, as well as in control areas (Jos City and Akure) (Arogunjo et al. 2009). High-purity germanium detectors (HPGe) were used for this analysis. Additionally, high-resolution sector field inductively coupled plasma-mass spectroscopy (HR-SF-ICP-MS) was employed to determine uranium and thorium levels in locally consumed liquids and foodstuffs from the mining area. Uranium and thorium activities in the soils and mineral sands of Bisichi ranged from 8.7 to 51 kBq/kg for 238U and from 16.8 to 98 kBq/kg for 232Th. These values were notably higher than those found in the control areas of Jos City and Akure, as well as exceeding reference values reported in the literature and concentrations in areas with high natural radioactive background. Concentrations of radionuclides in local foodstuffs and water samples from Bisichi exceeded UNSCEAR reference values, indicating the potential pollution impact of mining activities on the surrounding areas. John et al. (2020) measured the activity concentrations of natural radionuclides 238U and 232Th in granitic rock samples from uranium mine sites in Northern Nigeria using ICP-MS. The highest concentrations of 238U and 232Th (924.56 ± 17.13 and 21.96 ± 1.28 Bq/kg, respectively) were found at MikaI and Riruwai. The study also estimated and compared various radiological exposure parameters (absorbed dose rate in air, annual effective dose, radium equivalent activity, external exposure index, internal exposure index, and representative level index) to international recommended values. The study concluded that terrestrial gamma radiation from granitic rock in the study area does not pose significant radiation exposure to workers and residents.

Mwalongo et al. (2023) evaluated the uranium concentrations in four major phosphate rocks from East Africa and four commonly used mineral phosphate fertilizers in the region. Concentration measurements were conducted using energy-dispersive X-ray fluorescence spectrometry. The results indicated that uranium concentrations in phosphate rock varied from 10.7 mg/kg (Mrima Hill deposit, Kenya) to 631.6 mg/kg (Matongo deposit, Burundi). In phosphate fertilizers, uranium concentrations ranged from 107.9 mg/kg for an imported fertilizer to 281.0 mg/kg for a local fertilizer produced from Minjingu phosphate rock in Tanzania. It's worth noting that the natural concentration of uranium in the Earth's crust is typically between 1.4 and 2.7 mg/kg, while uranium mines in Namibia process ores with uranium concentrations as low as 100–400 mg/kg. The findings confirm that East African phosphate rock and its derived phosphate fertilizer can contain relatively high uranium concentrations. In Tanzania, Kasoga et al. (2015) detected elevated levels of natural radiation at the potential Manyoni uranium mines in Singida. Baseline data on ambient gamma radiation dose rates were collected to serve as a reference before, during, and after uranium mining activities. The absorbed gamma dose rates at the surface and one metre above the surface were measured using a survey meter, with sampling coordinates determined by GPS. The dose rates attributed to natural radionuclides ranged from 131 to 1678 nGy/h (98 to 1657 nGy/h) with a mean of 458 nGy/h (436 nGy/h). This maximum value was approximately 30 times higher than the global average of 59 nGy/h. The annual effective dose rates for the area varied from 0.16 to 2.06 mSv/y. High levels of gamma radiation were particularly noted in Mwanzi and Kinangali villages, indicating the necessity for further studies to assess health risks for the local population near the proposed uranium mining site.

2.6 FUTURE STUDIES

Radon and uranium exposure in communities proximal to mine tailings present significant health challenges, particularly in the Africa continent where there is a wealth of gold and uranium mines. Future research should focus on comprehensive longitudinal health studies to evaluate the long-term effects of radon and uranium exposure on populations living near mine tailing facilities. These studies should involve regular health check-ups and biomonitoring of individuals over extended periods, combined with detailed exposure assessments. Tracking health outcomes such as cancer incidence, respiratory diseases, and kidney dysfunction will provide crucial data on the chronic effects of low-dose radiation exposure from these mining activities. By establishing a clear link between exposure levels and health outcomes, policymakers can implement more effective health interventions and regulatory measures. Detailed geographic and temporal mapping of radon and uranium concentrations in soil, air, and water around mine tailing facilities is essential. Using advanced geospatial tools and satellite imagery, scholars can create high-resolution maps that indicate contamination hotspots and seasonal variations in exposure levels. Such maps would help identify the most affected areas and times of the year when exposure risks are highest. Additionally, integrating these maps with demographic data can aid in pinpointing vulnerable populations, leading to targeted public health interventions and more effective environmental monitoring programmes.

Investigating the mechanisms through which radon and uranium spread from mine tailings into the surrounding environment is crucial for developing mitigation strategies. Studies should focus on understanding the roles of wind patterns, water runoff, soil composition, and vegetation in the dispersion of these contaminants. By employing modelling techniques and field experiments, scholars can predict how changes in climate and land use might affect contaminant spread. This knowledge will be invaluable in designing engineering controls, such as barriers or vegetation buffers, to limit the migration of these hazardous substances. Furthermore, the development and testing of remediation techniques to reduce radon and uranium levels in contaminated areas should be explored. This could include phytoremediation, soil washing, and the use of adsorbent materials to capture and immobilize contaminants. The long-term efficacy, cost-effectiveness, and potential environmental impacts of these remediation strategies should also be explored. By comparing different techniques in varied environmental conditions across Africa, scholars can provide tailored recommendations for remediation practices that are most suitable for specific regions and contamination scenarios.

2.7 CONCLUSION

Investigating the socio-economic and policy implications of radon and uranium exposures from mine tailings is essential for crafting effective regulatory frameworks to complement African exposure scenarios. The economic costs of health care for affected populations, impact on local agriculture and livelihoods, and the potential benefits of implementing stricter environmental regulations remain unexplored options in the context of radon-uranium exposure studies within the continent. Moreover, engaging with communities through participatory research methods can provide insights into public perceptions of risk and the effectiveness of current communication strategies. This holistic approach will ensure that policies not only address the environmental and health impacts but also consider the socio-economic realities of the affected communities, leading to more sustainable and acceptable solutions.

REFERENCES

Abdul-Wahab, S. and Marikar, F., 2012. The environmental impact of gold mines: pollution by heavy metals. *Open Engineering*, 2(2), pp.304–313.

Abed, T.K., Fayad, M.A., Alamiery, A.A., Wahhab, H.A.A., Mohammed, J.K. and Majdi, H.S., 2024. Radon gas emission from home appliances: understanding sources, implications, and mitigation strategies. *Results in Engineering*, 1(22), p.102133.

Agius, R., Batistatou, E., Gittins, M., Jones, S., McNamee, R., Liu, H., Rashid, A., van Tongeren, M., Oertzen, G.V. and Wakeford, R., 2023. An epidemiological study of lung cancer and selected other cancers among Namibian Uranium workers. *Radiation Research*, 200(4), pp.340–348.

Alam, M.S. and Cheng, T., 2014. Uranium release from sediment to groundwater: influence of water chemistry and insights into release mechanisms. *Journal of Contaminant Hydrology*, 164, pp.72–87.

Anawar, H.M., 2015. Sustainable rehabilitation of mining waste and acid mine drainage using geochemistry, mine type, mineralogy, texture, ore extraction and climate knowledge. *Journal of Environmental Management*, 158, pp.111–121.

Arogunjo, A.M., Höllriegl, V., Giussani, A., Leopold, K., Gerstmann, U., Veronese, I. and Oeh, U., 2009. Uranium and thorium in soils, mineral sands, water and food samples in a tin mining area in Nigeria with elevated activity. *Journal of Environmental Radioactivity*, 100(3), pp.232–240.

Besada, H. and Martin, P., 2015. Mining codes in Africa: emergence of a 'fourth'generation?. *Cambridge Review of International Affairs*, 28(2), pp.263–282.

Bezuidenhout, J., 2019. Estimation of radon potential through measurement of uranium concentrations in granite geology. *South African Journal of Science*, 115(7–8), pp.1–4.

Brugge, D., deLemos, J.L. and Oldmixon, B., 2005. Exposure pathways and health effects associated with chemical and radiological toxicity of natural uranium: a review. *Reviews on Environmental Health*, 20(3), pp.177–194.

Bulut, H.A. and Şahin, R., 2024. Radon, concrete, buildings and human health—a review study. *Buildings*, 14(2), p.510.

Campbell, B. ed., 2009. *Mining in Africa: regulation and development*. Pluto Press, New York: IDRC.

Cherouati, D.E. and Djeffal, S., 1988. Measurements of radon and radon daughters in dwellings in Algiers. *Radiation Protection Dosimetry*, 25(2), pp.137–139.

Degu Belete, G. and Alemu Anteneh, Y., 2021. General overview of radon studies in health hazard perspectives. *Journal of Oncology*, 2021(1), p.6659795.

Dehkordi, A.N., 2012. Assessment of radiation doses to the West Rand public due to inhalation of 222Rn and its daughter products. *Faculty of Science*, p.112. (Master's thesis, University of the Witwatersrand, Johannesburg (South Africa)).

Dieu Souffit, G., Jacob Valdes, M., Bobbo Modibo, O., Flore, T.S.Y., Ateba Jean Félix, B. and Tokonami, S., 2022. Radon risk assessment and correlation study of indoor radon, radium-226, and radon in soil at the cobalt–nickel bearing area of Lomié, Eastern Cameroon. *Water, Air, & Soil Pollution*, 233(6), p.196.

Eggers, M.J., Doyle, J.T., Lefthand, M.J., Young, S.L., Moore-Nall, A.L., Kindness, L., Other Medicine, R., Ford, T.E., Dietrich, E., Parker, A.E. and Hoover, J.H., 2018. Community engaged cumulative risk assessment of exposure to inorganic well water contaminants, Crow Reservation, Montana. *International Journal of Environmental Research and Public Health*, 15(1), p.76.

Garg, V.K., Yadav, A., Singh, K., Singh, M., Bishnoi, M. and Pulhani, V., 2014. Uranium concentration in groundwater in Hisar city, India. *The International Journal of Occupational and Environmental Medicine*, 5(2), p.112.

Giri, S. and Singh, A.K., 2017. Ecological and human health risk assessment of agricultural soils based on heavy metals in mining areas of Singhbhum copper belt, India. *Human and Ecological Risk Assessment: An International Journal*, 23(5), pp.1008–1027.

Gümüşsoy, A., Başyiğit, M. and Kart, E.U., 2023. Economic potential and environmental impact of metal recovery from copper slag flotation tailings. *Resources Policy*, 80, p.103232.

Haakonde, T., Yabe, J., Choongo, K., Chongwe, G. and Islam, M.S., 2020. Preliminary assessment of uranium contamination in drinking water sources near a uranium mine in the Siavonga District, Zambia, and associated health risks. *Mine Water and the Environment*, 39(4), pp.735–745.

Haakonde, T., Yabe, J., Choongo, K., Chongwe, G. and Nchima, G., 2021. Uranium exposure and health risk implications: a preliminary study among the residents living around uranium mining sites in the Southern Province of Zambia. *Environmental Advances*, 5, p.100098.

Höllriegl, V., Arogunjo, A.M., Giussani, A., Michalke, B. and Oeh, U., 2011. Daily urinary excretion of uranium in members of the public of Southwest Nigeria. *Science of the Total Environment*, 412, pp.344–350.

International Atomic Energy Agency (IAEA)., 2019. Design and conduct of indoor radon surveys; Safety Reports Series No. 98, Austria.

International Commission on Radiological Protection (ICRP)., 1994. Protection against radon-222 at home and at work. ICRP Publ., 65. Ann ICRP 23: 1–38.

John, S.O.O., Usman, I.T., Akpa, T.C., Abubakar, S.A. and Ekong, G.B., 2020. Natural radionuclides in rock and radiation exposure index from uranium mine sites in parts of Northern Nigeria. *Radiation Protection and Environment*, 43(1), pp.36–43.

Johnson, C.A., 2015. The fate of cyanide in leach wastes at gold mines: an environmental perspective. *Applied Geochemistry*, 57, pp.194–205.

Kamunda, C., Mathuthu, M. and Madhuku, M.J.J.E.T.S., 2017. Determination of Radon in mine dwellings of Gauteng province of South Africa using AlphaGUARD radon professional monitor. *Journal of Environmental and Toxicological Studies*, 1, pp.1–4.

Kasoga, K.F., Mwalongo, D.A., Sawe, S.F., Nyaruba, M.M. and Dammalapati, U., 2015. Ambient gamma dose rate measurements at Manyoni uranium deposits, Singida, Tanzania. *Proceedings of SA Institute of Physics* (pp.180–185).

Kgabi, N.A., Mokgethi, S.A. and Bubu, A., 2009. Atmospheric radioactivity associated with gold mining activities in the North West Province of South Africa. *European Journal of Scientific Research*, 33(4), pp.594–605.

Khattak, N.U., Khan, M.A., Ali, N. and Abbas, S.M., 2011. Radon Monitoring for geological exploration: a review. *Journal of Himalayan Earth Sciences*, 44(2), pp.91–102.

Kneen, M.A., 2014. *Gold mine dumps in Witwatersrand and the associated health risk to exposed population*. The University of Texas at Dallas.

Komati, F., Ntwaeaborwa, M. and Strydom, R., 2022. An In Toto approach to Radon dispersion modelling from a South African gold mine tailings. *International Journal of Environmental Research and Public Health*, 19(13), p.8201.

Kumar, A., Rout, S., Narayanan, U., Mishra, M.K., Tripathi, R.M., Singh, J., Kumar, S. and Kushwaha, H.S., 2011. Geochemical modelling of uranium speciation in the subsurface aquatic environment of Punjab State in India. *Journal of Geology and Mining Research*, 3(5), pp.137–146.

Kumar, S., Kar, A.S., Rawat, N., Maity, S., Mishra, S., Pandit, G.G. and Tomar, B.S., 2015. Distribution coefficients of radionuclides around uranium mining area and effect of different analytical parameters on their determination. *Journal of Radioanalytical and Nuclear Chemistry*, 304, pp.727–733.

Le Roux, R.R., Bezuidenhout, J. and Smit, H.A.P., 2019. The influence of different types of granite on indoor radon concentrations of dwellings in the South African West Coast Peninsula. *Journal of Radiation Research and Applied Sciences*, 12(1), pp.375–382.

Leuchner, A. H., van As, D., Grundling, A. and Steyn, A., 1989. A survey of indoor exposure to radon in South Africa. *Clean Air Journal*, 7(7), pp.3–5.

Liefferink, M., 2019. "Selected extracts from South Africa's environmental legislation: challenges with the management of gold tailings within the Witwatersrand gold fields and case studies." In Paste 2019: Proceedings of the 22nd International Conference on Paste, Thickened and Filtered Tailings (pp. 53–67). Australian Centre for Geomechanics.

Lindsay, R., Newman, R.T. and Speelman, W.J., 2008. A study of airborne radon levels in Paarl houses (South Africa) and associated source terms, using electret ion chambers and gamma-ray spectrometry. *Applied Radiation and Isotopes*, 66(11), pp.1611–1614.

Liu, Y., Xu, Y., Xu, W., He, Z., Fu, C. and Du, F., 2024. Radon and lung cancer: current status and future prospects. *Critical Reviews in Oncology/Hematology*, 198(1), p.104363.

Ma, M., Wang, R., Xu, L., Xu, M. and Liu, S., 2020. Emerging health risks and underlying toxicological mechanisms of uranium contamination: lessons from the past two decades. *Environment International*, 145, p.106107.

Maseki, J., 2013. *Risk assessment of inhaled and ingested airborne particles in the vicinity of gold mine tailings: case study of the Witwatersrand Basin*. University of Johannesburg (South Africa).

Maseki, J., Annegarn, H.J. and Spiers, G., 2017. Health risk posed by enriched heavy metals (As, Cd, and Cr) in airborne particles from Witwatersrand gold tailings. *Journal of the Southern African Institute of Mining and Metallurgy*, 117(7), pp.663–669.

Mbida Mbembe, S., Akamba Mbembe, B., Manga, A., Ele Abiama, P., Saidou, Ondo Meye, P., Owono Ateba, P. and Ben–Bolie, G.H., 2023. Indoor radon, external dose rate and assessment of lung cancer risk in dwellings: the case of Ebolowa town, Southern Cameroon. *International Journal of Environmental Analytical Chemistry*, 103(17), pp.5957–5973.

Moshupya, P., Abiye, T., Mouri, H., Levin, M., Strauss, M. and Strydom, R., 2019. Assessment of radon concentration and impact on human Health in a region dominated by abandoned gold mine tailings dams: a case from the west Rand region, South Africa. *Geosciences*, 9(11), p.466.

Moshupya, P.M., Mohuba, S.C., Abiye, T.A. and Korir, I., 2023. Evaluation of indoor radon activity concentrations and controls in dwellings surrounding the gold mine tailings in Gauteng Province of South Africa. *International Journal of Environmental Research and Public Health*, 20(21), p.7010.

Mphaga, K.V., Utembe, W. and Rathebe, P.C., 2024. Radon exposure risks among residents proximal to gold mine tailings in Gauteng Province, South Africa: a cross-sectional preliminary study protocol. *Frontiers in Public Health*, 12, p.1328955.

Mphaga, K.V., Utembe, W., Shezi, B., Mbonane, T.P. and Rathebe, P.C., 2024. Unintended consequences of urban expansion and gold mining: elevated indoor radon levels in Gauteng communities' neighboring gold mine tailings. *Atmosphere*, 15(8), p.881.

Mwalongo, D.A., Haneklaus, N.H., Lisuma, J.B., Kivevele, T.T. and Mtei, K.M., 2023. Uranium in phosphate rocks and mineral fertilizers applied to agricultural soils in East Africa. *Environmental Science and Pollution Research*, 30(12), pp.33898–33906.

Ndjana Nkoulou II, J.E., Manga, A., Saïdou, German, O., Sainz-Fernandez, C. and Kwato Njock, M.G., 2022. Natural radioactivity in building materials, indoor radon measurements, and assessment of the associated risk indicators in some localities of the Centre Region, Cameroon. *Environmental Science and Pollution Research*, 29(36), pp.54842–54854.

Ngole-Jeme, V.M. and Fantke, P., 2017. Ecological and human health risks associated with abandoned gold mine tailings contaminated soil. *PloS one*, 12(2), p.e0172517.

Njinga, R.L., Tshivhase, V.M. and Mathuthu, M., 2016. Chemical toxicity of surface-based drinking water sources due to natural uranium pollutant around princess gold mine environs in Roodepoort, South Africa. *Exposure and Health*, 8, pp.457–464.

Nriagu, J., Nam, D.H., Ayanwola, T.A., Dinh, H., Erdenechimeg, E., Ochir, C. and Bolormaa, T.A., 2012. High levels of uranium in groundwater of Ulaanbaatar, Mongolia. *Science of the Total Environment*, 414, pp.722–726.

Obed, R.I., Ademola, A.K. and Ogundare, F.O., 2012. Radon measurements by nuclear track detectors in dwellings in Oke-Ogun area, South-Western, Nigeria. *Radiation Protection Dosimetry*, 148(4), pp.475–481.

Oberhauser, A.M., 2016. (Re) constructing rural–urban spaces: gendered livelihoods, migration, and natural resources in South Africa. *GeoJournal*, 81, pp.489–502.

Ocwelwang, A., Sethabela, C., Mathuthu, M. and Moshupya, P., 2019. Indoor Radon research conducted in South Africa from 1980s–to date (2019): a review. https://irpa.net/irpa15/T7.1-O0181.pdf

Owen, J.R., Kemp, D., Lébre, É., Svobodova, K. and Murillo, G.P., 2020. Catastrophic tailings dam failures and disaster risk disclosure. *International Journal of Disaster Risk Reduction*, 42, p.101361.

Pashkevich, M.A., 2017. Classification and environmental impact of mine dumps. In *Assessment, restoration and reclamation of mining influenced soils* (pp. 1–32). Academic Press, Candice Janco, United Kingdom.

Raji, I.B., 2023. The dynamics of fluvial transportation of uranium in the Rietspruit river system in South Africa (Doctoral dissertation, North-West University (South Africa)).

Ramadan, R.S., Dawood, Y.H., Yehia, M.M. and Gad, A., 2022. Environmental and health impact of current uranium mining activities in southwestern Sinai, Egypt. *Environmental Earth Sciences*, 81(7), p.213.

Rehman, K., Fatima, F., Waheed, I. and Akash, M.S.H., 2018. Prevalence of exposure of heavy metals and their impact on health consequences. *Journal of Cellular Biochemistry*, 119(1), pp.157–184.

Riudavets, M., Garcia de Herreros, M., Besse, B. and Mezquita, L., 2022. Radon and lung cancer: current trends and future perspectives. *Cancers*, 14(13), p.3142.

Robertson, A., Allen, J., Laney, R. and Curnow, A., 2013. The cellular and molecular carcinogenic effects of radon exposure: a review. *International Journal of Molecular Sciences*, 14(7), pp.14024–14063.

Sahu, P., Panigrahi, D.C. and Mishra, D.P., 2016. A comprehensive review on sources of radon and factors affecting radon concentration in underground uranium mines. *Environmental Earth Sciences*, 75, pp.1–19.

Salupeto-Dembo, J., Szabó-Krausz, Z., Völgyesi, P. and Szabó, C., 2020. Radon and thoron radiation exposure of the Angolan population living in adobe houses. *Journal of Radioanalytical and Nuclear Chemistry*, 325, pp.271–282.

Schmid, K., Kuwert, T. and Drexler, H., 2010. Radon in indoor spaces: an underestimated risk factor for lung cancer in environmental medicine. *Deutsches Arzteblatt International*, 107(11), p.181.

Seldén, A.I., Lundholm, C., Edlund, B., Högdahl, C., Ek, B.M., Bergström, B.E. and Ohlson, C.G., 2009. Nephrotoxicity of uranium in drinking water from private drilled wells. *Environmental Research*, 109(4), pp.486–494.

Setso, R., 2020. Assessment of indoor radon concentration levels in Serule, Botswana (Master's thesis, Botswana International University of Science & Technology (Botswana)).

Shanmukha, N.T., Vinayaka, M., Lokeshappa, B. and Nadaf, S., 2024. Biodiversity loss due to mining activities. In *Impact of societal development and infrastructure on biodiversity decline* (pp. 166–191). IGI Global.

Sibanda, L.K., 2019. *Repurposing of mine waste: an alternative management approach to gold tailings in South Africa*. University of Cape Town, Doctoral dissertation.

Singh, L., Kumar, R., Kumar, S., Bajwa, B.S. and Singh, S., 2013. Health risk assessments due to uranium contamination of drinking water in Bathinda region, Punjab state, India. *Radioprotection*, 48(2), pp.191–202.

Sonone, S.S., Jadhav, S., Sankhla, M.S. and Kumar, R., 2020. Water contamination by heavy metals and their toxic effect on aquaculture and human health through food Chain. *Letters in Applied NanoBioScience*, 10(2), pp.2148–2166.

Srivastava, R.R., Pathak, P. and Perween, M., 2020. Environmental and health impact due to uranium mining. In: Gupta, D. and Walther, C. (eds), *Uranium in Plants and the Environment*, pp.69–89. Springer, Cham.

Starościak, E. and Rosiak, L., 2015. Determination of uranium reference levels in the urine of Warsaw residents (Poland). *Journal of Radioanalytical and Nuclear Chemistry*, 304, pp.75–79.

Sundal, A.V., Henriksen, H., Soldal, O. and Strand, T., 2004. The influence of geological factors on indoor radon concentrations in Norway. *Science of the Total Environment*, 328(1–3), pp.41–53.

Tanwer, N., Arora, V., Kant, K., Singh, B., Laura, J.S. and Khosla, B., 2024. Prevalence of Uranium in groundwater of rural and urban regions of India. In *Water resources management for rural development* (pp. 213–234). Elsevier, United Kingdom.

Timsina, S., Hardy, N.G., Woodbury, D.J., Ashton, M.S., Cook-Patton, S.C., Pasternack, R. and Martin, M.P., 2022. Tropical surface gold mining: a review of ecological impacts and restoration strategies. *Land Degradation & Development*, 33(18), pp.3661–3674.

Tsewu, A., 2021. Radioactivity assessment of radium and radon concentrations in water sources at and near selected former uranium mines in the west-rand area of Johannesburg (Doctoral dissertation, University of the Witwatersrand).

Voldřich, A., 2019. Radon, the silent killer: the neglected repercussions cause unforeseen environmental health risks in African countries (Doctoral dissertation, Wien).

Winde, F., Brugge, D., Nidecker, A. and Ruegg, U., 2017. Uranium from Africa–an overview on past and current mining activities: re-appraising associated risks and chances in a global context. *Journal of African Earth Sciences*, 129, pp.759–778.

Winde, F., Hoffmann, E., Espina, C. and Schüz, J., 2019. Mapping and modelling human exposure to uraniferous mine waste using a GIS-supported virtual geographic environment. *Journal of Geochemical Exploration*, 204, pp.167–180.

World Health Organization., 2001. *Depleted uranium: sources, exposure and health effects* (No. WHO/SDE/PHE/01.1). World Health Organization.

World Health Organization., 2009. *WHO handbook on indoor radon: a public health perspective*. Geneva: World Health Organization.

Wymer, D.G. and Botha, J.C., 2001, September. Managing the environmental impacts of low activity wastes from the South African gold mining industry. In International Conference on Radioactive Waste Management and Environmental Remediation (Vol. 80180, pp.1435–1440). American Society of Mechanical Engineers.

Zhang, C., Wang, X., Jiang, S., Zhou, M., Li, F., Bi, X., Xie, S. and Liu, J., 2021. Heavy metal pollution caused by cyanide gold leaching: a case study of gold tailings in central China. *Environmental Science and Pollution Research*, 28, pp.29231–29240.

Zupunski, L., Street, R., Ostroumova, E., Winde, F., Sachs, S., Geipel, G., Nkosi, V., Bouaoun, L., Haman, T., Schüz, J. and Mathee, A., 2023. Environmental exposure to uranium in a population living in close proximity to gold mine tailings in South Africa. *Journal of Trace Elements in Medicine and Biology*, 77, p.127141.

3 Non-ionizing radiation exposure assessments for communication and power sources

3.1 BACKGROUND OF THE CHAPTER

Non-ionizing radiation (NIR) has become an integral aspect of modern society, emanating from a variety of communication and electric power sources. Unlike ionizing radiation, which carries enough energy to ionize atoms and potentially cause cellular damage, NIR is generally considered less harmful. However, the pervasive exposure to NIR from devices such as mobile phones, Wi-Fi routers, power lines, and other electrical appliances has raised public health concerns and scientific interest (Ndoma et al. 2024).

In the realm of communication, the proliferation of wireless technologies has led to increased exposure to radiofrequency (RF) radiation. Mobile phones, in particular, are ubiquitous, and their proximity to the body during use has sparked extensive research into potential health effects. The International Agency for Research on Cancer (IARC) has classified RF radiation as possibly carcinogenic to humans (Group 2B), citing a potential increased risk for glioma, a malignant type of brain cancer (Gupta et al. 2022). Several epidemiological studies have examined the correlation between long-term mobile phone use and the incidence of brain tumours, with mixed results. For instance, the INTERPHONE study, one of the largest case-control studies on this topic, found an increased risk of glioma among heavy users of mobile phones, though the findings were not conclusive (INTERPHONE Study Group 2010).

Similarly, Wi-Fi technology, which operates in the RF spectrum, is prevalent in homes, schools, and workplaces, contributing to ambient exposure levels. Research has investigated the potential biological effects of Wi-Fi exposure, including impacts on cognitive functions and reproductive health. A study by Aydin and Cagiltay (2012) highlighted that while some studies suggest possible adverse effects, the evidence remains inconsistent and further research is needed to establish definitive conclusions.

Electric power sources, however, primarily emit extremely low-frequency electromagnetic fields (ELF-EMFs). These fields are generated by power lines, electrical wiring, and household appliances. The widespread distribution of electrical infrastructure means that ELF-EMFs are a constant presence in daily life, prompting investigations into their long-term health implications. The World Health Organization (WHO) has noted that while ELF-EMFs are not known to cause

DOI: 10.1201/9781003508588-3

TABLE 3.1
NIR and associated exposure symptoms

NIR	Health-induced outcome	References
Radiofrequency	The potency of the spermatozoa (risk of infertility), nerve stimulation, changes in the permeability of cell membranes	Otitoloju et al. (2010); Oyewopo et al. (2017)
Microwaves	Sleep disorders, physiological effects, and learning and memory impairments	Banik et al. (2003); Zhi et al. (2017)
Infrared rays	Premature skin-ageing/photo-ageing	Schieke et al. (2003); Cho et al. (2009)
Visible light	Photo-ageing, Erythema	Pourang et al. (2022); Sklar et al. (2013); Guan et al. (2021)
Ultraviolet radiation	Skin Cancers (In 2020, 1.5 million cases of skin cancers were diagnosed, and 120,000 skin cancer-associated deaths were reported), DNA damage, and immune system suppression	Gallagher et al. (2010)

significant health issues at low exposure levels, there is limited evidence suggesting a possible association with childhood leukaemia at higher exposure levels (WHO 2007). Several epidemiological studies have explored the link between residential proximity to high-voltage power lines and an increased risk of leukaemia in children. For example, a pooled analysis by Kheifets and Oksuzyan (2008) found a small but statistically significant association between ELF-EMF exposure and childhood leukaemia. However, the authors cautioned that the results could be influenced by confounding factors and emphasized the need for more rigorous studies to confirm these findings.

Understanding the mechanisms of NIR is crucial for evaluating its potential health effects. Table 3.1 provides an overview of exposure effects following exposure to NIR. RF radiation primarily interacts with biological tissues through thermal effects, causing heating of tissues at high exposure levels. However, at the levels emitted by communication devices, the thermal effects are minimal, leading researchers to investigate non-thermal mechanisms of action. A review by Foster and Repacholi (2004) discussed various hypotheses, including oxidative stress and changes in cell membrane function, though conclusive evidence is still lacking. ELF-EMFs, due to their lower frequencies, interact with biological tissues differently. They can induce electric currents within the body, which may influence cellular processes. Research by Kheifets et al. (2005) examined the potential mechanisms by which ELF-EMFs could contribute to carcinogenesis, suggesting that prolonged exposure might interfere with cell cycle regulation and DNA repair processes.

Regulatory bodies have established guidelines to limit NIR exposure and protect public health. The International Commission on Non-Ionizing Radiation Protection (ICNIRP) provides exposure limits for RF and ELF-EMFs based on a thorough review of scientific evidence. These guidelines aim to prevent known adverse health effects by maintaining exposure levels well below thresholds where significant biological interactions occur (ICNIRP 2020). However, given the evolving nature of NIR research, ongoing updates to these guidelines are essential to incorporate new findings and ensure adequate protection.

3.2 OVERVIEW OF NON-IONIZING RADIATION EXPOSURE ASSESSMENTS FROM COMMUNICATION DEVICES IN AFRICA

Evaluating NIR exposure levels in both urban and rural settings is critical due to the widespread use of wireless communication and related devices that operate on radiofrequency (RF) energies. High levels of NIR exposure have been associated with adverse health effects, such as nerve stimulation and alterations in cell membrane permeability. The ICNIRP sets guidelines for NIR exposure for both workers and the general population. Chege (2015) conducted an exposure assessment study in Nairobi, Kenya's capital. The electromagnetic frequency fields (EMF) were systematically measured using a spectrum analyser (Spectran HF6065) connected to a HyperLOG 7060 antenna in the Central Business District (CBD) and three densely populated areas on its outskirts. The results were further compared to the ICNIRP standards. The median EMF exposure values in Nairobi's CBD and residential areas were 0.90 ± 0.30 V/m and 0.81 ± 0.25 V/m, respectively, with total exposure values ranging from 0.14 V/m to 3.67 V/m. GSM900 was identified as the frequency band contributing the highest NIR exposure compared to LTE700, LTE800, GSM1800, UMTS2100, and LTE2600. Utilizing the Kriging technique, a spatial map was created to illustrate the distribution of EMF measurements, highlighting areas in the CBD with a high probability of EMF levels exceeding 2 V/m. The results indicated that EMF exposure in all assessed areas was below ICNIRP guideline limits. In Nigeria, health risks associated with NIR from telecommunication masts on the local population were evaluated through a descriptive cross-sectional survey. Socio-demographic data and hazard profiles of the respondents were recorded. The hazard profile revealed that most respondents (62%) experienced various symptoms, with headaches being the most common (51.6%), aligning with established research findings. A significant synergistic relationship was found between high-voltage cables and telecommunication masts concerning health effects, with a p-value <0.05. Additionally, it was demonstrated that the proximity to and duration of exposure to mast radiation were directly proportional to the observed health hazards (p-value <0.05).

In South Africa, Rathebe (2023) evaluated radiofrequency electromagnetic fields (RF-EMF) exposure levels near primary schools in Mpumalanga province. Using a calibrated Acoustimeter, RF-EMF levels were measured at distances of 50 and 100 m from three mobile base stations providing 4G coverage. Measurements were taken from each of the three stations, representing different district municipalities, in 30-minute intervals. The peak exposure levels were recorded as 887 $\mu W/m^2$ at 50 metres and 905 $\mu W/m^2$ at 100 m. No significant statistical differences were observed when comparing the data from the three municipalities at 50 and 100 m. However, measurements comparing all base stations at 50 ($p < 0.002$) and 100 metres ($p < 0.003$) were statistically significant. The RF-EMF exposure levels at both distances were below the reference levels set by the ICNIRP. In the same country, the RF-EMF exposure levels were measured in two unoccupied residential apartments near a network base station in a peri-urban setting (Rathebe 2022). RF-EMF levels were measured using a TM-196 3 Axis RF Field strength meter in six different rooms of the buildings at various times: 8 AM, 3 PM, and 7:30 PM. The highest exposure value recorded was 9842 μW/m² in the lounge area of residential building 2, while the lowest was 15.5 μW/m² on the balcony of building 1. ANOVA indicated no significant statistical difference when comparing measurements taken at different times ($p < 0.94$ and $p < 0.54$) or between the two buildings ($p < 0.50$). The lower RF-EMF levels were attributed to the reflective building structures, as no vegetation was present during the measurement planning period. The peak RF levels were linked to increased network activities in the nearby residential house. All measured RF levels were below the reference levels specified in the ICNIRP RF guidelines.

Ajibare et al. (2023) examined the specific absorption rate (SAR) and exposure index (EI) of access points (APs) and user equipment (UEs) in fourth generation (4G) and fifth generation (5G) wireless technologies, focusing on the effects of exposure to RF-EMFs and strategies for reducing this exposure. The EI was characterized using a classical mathematical approach that considers power density, SAR, electric field strength, and the tissue's density and conductivity. A novel exposure-index open-loop power control algorithm was proposed to evaluate realistic RF-EMF radiation exposure on human users from both downlink (DL) and uplink (UL) communication devices. To address the EI minimization problem using the open-loop power control algorithm, a mixed-integer nonlinear programming (MINLP) problem was formulated. Since energy capacity (i.e., power density) in wireless networks influences radiation exposure (SAR and EI), the approach minimized the EI by managing transmitted and received power while adhering to Quality of Service (QoS), interference, and power constraints, ensuring that users' QoS requirements are met. The proposed scheme is numerically compared with other heuristic algorithms and exposure limits set by the ICNIRP and similar organizations. Additionally, simulation results indicated that the proposed technique is an effective alternative, supported by numerical results and evaluation, confirms that the exposures are within recommended limits.

Inyang et al. (2017) determined the RF-EMFs emitted by Wi-Fi routers at varying distances from 0.5 m to 3 m using an electrosmog meter, in Nigeria. The highest RF levels recorded were 1.600 V/m, 1.569 V/m, 1.450 V/m, 1.346 V/m, 1.237 V/m, 0.994 V/m, and 0.578 V/m, all at a distance of 0.5 m across various sample locations. These measurements were significantly below the exposure limits established by the ICNIRP. While short-term exposure did not show any health effects, long-term exposure, proximity to the radiation source, and the strength of the electromagnetic field generated can potentially pose health risks. In Malawi, Blantyre, Nahuku et al. (2018) conducted a descriptive study and employed a Spectran HF V4 spectrum analyser to measure RF-EMF levels at 25 m intervals from the fence of the base transceiver stations (BTS) to a maximum radial distance of 150 m. The city recorded a maximum power density of 0.00422 W/m² at BTS 5. All recorded values were below the ICNIRP standard guidelines of 1–10 W/m², with far fields reaching the ground at approximately 30 m to 150 m. Maduka et al. (2019) measured the power densities from various telecommunication masts operated by different network providers using a spectrum analyser. RF-EMF levels were measured within a 100 m radius around selected areas in the Gusau local government area. The highest mean power densities recorded for MTN, Airtel, and 9mobile were 45.60 nW/m^2, 22.08 nW/m^2, and 34.56 nW/m^2, respectively, while the lowest values were 11.59 nW/m^2, 18.62 nW/m^2, and 33.42 nW/m^2, respectively. The fluctuations in values were influenced by factors such as wave interference from other electromagnetic sources near the reference base stations. The study compared the measured mean power densities from individual mobile transceiver stations (MTN, Airtel, and 9mobile) with the standard limit set by the ICNIRP. The measured radiation exposure levels were below the standard limit of 4.5 W/m^2 for a 900 MHz system, indicating that the exposure levels in these areas are low and do not pose significant health risks to the residents.

A study by Garba et al. (2017) assessed RF-EMFs from selected BTSs operated by different network providers in Kaduna State, Nigeria, using a B and K Precision model 2658A spectrum analyser. RF levels were assessed within a 100 m radius around certain areas in Kaduna State. The measured values ranged from 9.29 nW/m^2 to 58.08 nW/m^2. The results indicated that the radiation exposure levels were below the standard limit of 4.5 W/m^2 for a 900 MHz system set by the ICNIRP. In Nigerian, RF power density measurements were conducted to assess exposure levels at public locations. Data on power density were gathered from base stations (BSs) operated by three major GSM operators in Nigeria, all located in GRA, Benin City. Measurements were taken at 5-m intervals up to 100 m from each BS. The highest recorded power densities for GLo, MTN, and Airtel were 1.62 mW/m^2, 3.04 mW/m^2, and 1.52 mW/m^2, respectively. These values are significantly lower than international standard limits (Ojuh and Isabona 2015). An electrosmog meter was utilized to measure RF power densities from 12 BTS located across four local governments in Ogbomoso, Nigeria at various distances. The measurements aimed to assess the mean power densities throughout Ogbomoso and its surroundings against the standards set by the ICNIRP. The highest average power density recorded was 718.6 μW/m^2 at a distance of 25 m from the mast, which falls below the ICNIRP guidelines (Annida et al. 2021). Also, in Dutsin-Ma, Katsina State, Nwamaka et al. (2023) determined the RF levels proximal to various mobile phones. Real-time measurements of power densities were conducted on 50 different mobile phones of varying sizes, types, and models at different distances using a Cornet Electrosmog RF Meter. Measurements were taken both in standby mode and during active use, with the RF meter positioned at intervals ranging from 10 cm to a maximum of 50 cm from each phone. The average power densities observed during active use ranged from 0.799 ± 0.11 mW/m^2 to 182.700 ± 10.91 mW/m^2, all of which were found to be below the safety limits.

Olorunsola et al. (2021) investigated the temporal variability of RF field exposure from selected Mobile Communication Base Stations (MCBSs) in Kuje Area Council, Nigeria. The study focused on mobile phone frequency bands including GSM 900 MHz, GSM 1800 MHz, and Wi-Fi 2400 MHz. Measurements of peak power densities and their corresponding times were conducted using a selective radiation spectral analyser HF-2025E. The results demonstrated that RF fields exhibit temporal variations across different locations, resulting in varying power densities over time. Comparisons of the maximum exposures aggregated from these measurements with the guidelines set by the ICNIRP indicated that they remained below the recommended limits. In Kenya, Wilson et al. (2011) found RF levels from all tested mobile phone models ranged from 0.01134 to 0.4671 $mWcm^2$, with the highest recorded from Nokia Series (China) N95 and the lowest from Nokia 1110. However, all were below the recommended exposure limits. Deatanyah et al. (2020) assessed RF field levels in residential areas, schools, and marketplaces of Ghana to compare the exposure levels against the ICNIRP guidelines. Utilizing a calibrated log-periodic antenna paired with a spectrum analyser, measurements were conducted. The recorded results ranged from 0.0008 to 182 mW m^2, demonstrating overall adherence to the ICNIRP reference levels. The highest recorded value represented 1.39% of the reference limit.

In Sudan, Mohammed et al. (2022) classified and characterized RF-EMF exposure from mobile phone BSs in Khartoum city. Spatial distribution measurements were conducted during two periods: in 2012 and again from September 2019 to January 2020, encompassing 282 antennas operating with GSM900, GSM1800, and UMTS2100 frequencies. The antennas were associated with three mobile communication companies; Sudani, Zain, and MTN. Measurements were performed using a frequency-selective RF analyser at fixed distances from the antennas. Average measurements for GSM900, GSM1800, and UMTS were recorded as 0.01933 W/m^2, 0.0067 W/m^2, and 0.0046 W/m^2, respectively. Peak power densities were notably observed at distances of 90 m, 110 m, 130 m, and 150 m for the majority (70%) of the antennas, with the highest values typically occurring at 110 m. Visual representation of group centroids supported the accuracy of frequency-selective measurements, confirming distinct RF signals from different companies. Statistical analyses confirmed significant differences between the companies for UMTS and GSM900 measurements, but not for GSM1800. Kriging interpolation using ArcGIS revealed a notable spatial distribution of RF exposure across the study area, with marketplaces and typical urban residential areas exhibiting the highest levels, exceeding ICNIRP limits.

Thulu et al. (2023) measured electric field and magnetic flux density levels in Blantyre City, Malawi, between 2020 and 2021. Sixty short-term measurements were conducted using the TriField Electro Magnetic Frequency meter model TF2 across 30 different locations. Sampling points included high-population-dense areas such as school campuses, hospitals, industrial zones, markets, residential areas, and the commercial and business centre (CBC) of Blantyre. Measurements were conducted during two time periods: 10:00–12:00h and 17:00–19:00h for short-range analysis. The maximum electric field levels recorded during these periods were 249.24 mV/m and 207.85 mV/m, respectively, both below the public exposure limit of 4200.00 mV/m. Similarly, the maximum magnetic flux levels were 0.073 G and 0.057 G during the same time slots, well below the public exposure limit of 2G. Joyner et al. (2014) analysed 260,000 measurement points from surveys of RF-EMFs strengths near radio BSs in seven African countries between 2001 to 2003 and 2006 to 2012. Summary of results are presented in Table 3.2.

The comprehensive assessment of NIR exposure from communication devices across various African regions highlights the generally low levels of exposure relative to international safety standards. Despite the rapid expansion of telecommunication infrastructure, including mobile BSs and Wi-Fi networks, the measured

TABLE 3.2
Overview of RF measurements from seven African countries between 2001 and 2021

Country	Period	Cellular technology	Number of measurements	Minimum (μW/cm²)	Maximum (μW/cm²)
Botswana	2010	GSM900	181	1.71E–04	4.92E+00
		GSM1800	181	9.20E–06	5.45E+00
		WCDMA	181	3.28E–05	1.47E+00
		2010 Totals	543	9.20E–06	5.45E+00
		FM radio	181	6.94E–05	1.92E+01
Ivory Coast	2009	GSM900	43	3.90E–03	4.47E+00
		GSM1800	43	3.32E–02	2.10E+01
	2010	GSM900	43	2.48E–03	5.62E+00
		GSM1800	43	1.31E–02	9.36E+00
		CDMA850	39	2.15E–05	2.99E–01
Ghana	2007	GSM900	50	1.00E–06	1.00E–03
		GSM1800	50	1.00E–06	1.00E–02
	2010/2011	GSM900	32	8.50E–02	1.07E–01
		GSM1800	42	7.80E–02	1.19E–01
Mauritania	2007	GSM900	146	1.53E–06	1.43E–01
		GSM1800	3	4.90E–06	9.14E–04
		CDMA	204	2.85E–03	2.51E–01
		2007 Totals	353	1.53E–06	2.51E–01
	2009	GSM900	108	7.65E–04	6.70E–01
		GSM1800	12	1.12E–02	7.21E–02
		CDMA	168	1.16E–03	5.25E–01
		2099 Totals	288	7.65E–04	6.70E–01
	2010	GSM900	130	8.68E–05	7.71E–01
		GSM1800	37	6.37E–04	3.40E–01
		CDMA	91	1.08E–04	1.30E–01
		2010 Totals	258	8.68E–05	7.71E–01
	2007–2010	2007–2010 Totals	899	1.53E–06	7.71E–01
Nigeria	2001–2003	GSM900/ GSM1800	212	—	5.94E–03
South Africa	2006	GSM900	21,016	3.60E–05	3.90E+02
		GSM1800	16,758	4.08E–05	9.94E+00
		WCDMA	14,673	5.48E–05	1.34E+02
		2006 Totals	52,447	3.60E–05	3.90E+02
		FM radio	930	2.74E–04	2.88E+00
	2007	GSM900	20,494	1.72E–05	1.50E+02
		GSM1800	15,771	1.84E–05	6.33E+01
		WCDMA	14,821	3.05E–05	2.47E+01
		2007 Totals	51,086	1.72E–05	1.50E+02
		FM radio	832	1.04E–04	4.52E+00

(Continued)

TABLE 3.2
Continued

Country	Period	Cellular technology	Number of measurements	Minimum (μW/cm²)	Maximum (μW/cm²)
	2008	GSM900	11,181	2.81E–05	1.58E+02
		GSM1800	7119	2.35E–05	5.41E+01
		WCDMA	6358	4.03E–05	9.64E+01
		2008 Totals	24,658	2.35E–05	1.58E+02
		FM radio	571	2.87E–04	3.25E+00
	2009	GSM900	6014	1.67E–05	1.59E+01
		GSM1800	4097	1.70E–05	2.25E+00
		WCDMA	3675	3.87E–05	1.74E+01
		2009 Totals	13,786	1.67E–05	1.74E+01
		FM radio	219	2.80E–04	9.13E+00
	2010	GSM900	4614	2.91E–05	4.77E+01
		GSM1800	2539	2.32E–05	4.70E+00
		WCDMA	3062	3.87E–05	1.16E+01
		2010 Totals	10,215	2.32E–05	4.77E+01
		FM radio	318	2.14E–04	1.88E+00
	2011	GSM900	3137	2.16E–05	3.85E+01
		GSM1800	2441	3.47E–06	5.69E+01
		WCDMA	2677	2.85E–05	3.79E+01
		2011 Totals	8255	3.47E–06	5.69E+01
		FM radio	88	2.75E–04	2.49E+00
	2012	GSM900	2589	2.13E–05	2.39E+02
		GSM1800	1855	1.79E–05	8.14E+01
		WCDMA	2078	3.29E–05	2.73E+01
		2012 Totals	6522	1.79E–05	2.39E+02
		FM radio	96	2.33E–04	1.76E+00
	2006–2012	GSM900	77,282	1.67E–05	3.90E+02
		GSM1800	56,955	3.47E–06	8.14E+01
		WCDMA	53,911	2.85E–05	1.34E+02
		2006–2012 Totals	188,148	3.47E–06	3.90E+02
		FM radio	3562	2.95E–05	9.13E+00
Zambia	2009	GSM 900	126	2.11E–04	5.29E–01
		GSM1800	64	5.29E–07	1.33E–01
		WCDMA	125	1.67E–04	1.06E–01
		Totals	315	5.29E–07	5.29E–01

radiation levels remain well within the limits set by the ICNIRP and other regulatory bodies. This consistency suggests that the current deployment of communication devices poses minimal health risks to the public. However, the findings also underscore the need for continuous monitoring and adherence to safety guidelines as technological advancements and infrastructure expansion continue. There is a need for exposure assessments in Africa to focus on long-term exposure effects and include diverse environmental settings to ensure comprehensive risk assessment.

Public awareness initiatives should also be undertaken to educate communities about NIR and its implications, thus fostering informed decision-making and safety practices.

3.3 OVERVIEW OF NON-IONIZING RADIATION EXPOSURE ASSESSMENTS FROM ELECTRIC POWER SOURCES IN AFRICA

NIR exposure assessments from electric power sources in Africa have become increasingly important due to the rapid expansion of electrical infrastructure across the continent. As countries develop and electrify rural and urban areas, understanding the levels and potential health impacts of exposure to ELF-EMFs generated by power lines, transformers, and other electrical devices is crucial. Despite the growing body of international research on ELF-EMFs, data specific to African contexts remain limited. This section aims to bridge that gap by examining exposure levels in various African regions, considering factors such as population density, infrastructure development, and seasonal variations. Assessing these exposure levels will provide valuable insights for public health policies and safety standards, ensuring that the benefits of electrification are realized without compromising community health.

In South Africa, Rathebe (2020) evaluated the exposure levels of extremely low-frequency magnetic fields (ELF-MFs) from medium overhead power lines in Ngodini, a rural township in the Mpumalanga province. Purposive random sampling was used to select 30 residential houses that were connected to electricity in 2015. A TriField model XE100 meter was utilized to measure emission data at heights of 0, 1, 1.5, and 1.8 m above ground level during both summer and winter. These heights were selected to represent various human organs' positions, including the feet, pelvic region, pleural region, and head. Measurements were taken in front of residential gates, areas of frequent resident movement. Significant differences were observed when comparing measurements at different heights ($p < 0.001$) and between the summer and winter seasons ($p < 0.0001$). Although the exposure levels were within the ICNIRP recommended limits, peaks of 2.87 μT in winter and 2.34 μT in summer were recorded. A case-study in Osogbo, Nigeria measured magnetic field component of the electromagnetic fields generated around transmission lines, using a gaussmeter at a vertical distance of 1.04 m at the centre, as well as the right and left sides, of selected power transmission lines (Badru et al. 2019). A buffer of 500 m was created along the power lines to measure the distance between the lines and the nearest infrastructure. The analysis of the magnetic field values recorded around a transmission tower revealed higher densities at the sides of the tower, particularly at right side, underneath, and left side. These values, although high, remain below the ICNIRP guideline. The spatial distance between power lines and the nearest infrastructure was often less than the required 15 m for 132 kV lines and 25 m for 330 kV lines, with many buildings and workplaces located directly below the power lines. In 2018, Rathebe et al. assessed the ELF-MF exposure levels in the central region of South Africa. Measurements were taken at 15 residential sites, using a TriField meter model XE 100. Data was collected at distances of 3 m, 6 m, and 9 m outside electrical substations, and at four different corners inside the substations, near barrier screening, designated as 0 m (reference point). The findings indicated no significant differences among the 15 residential areas. However, one residential area showed significantly higher exposure levels (0.55 μT) compared to other sites. The four distance intervals demonstrated a highly significant difference ($p < 0.0001$) when compared to each other. The t-test revealed statistically significant differences in exposure levels recorded at 3 m, 6 m, and 9 m compared to 0 m ($p < 0.01$). All recorded exposure levels were below ICNIRP guidelines and decreased rapidly with increased distance from the source.

ELF-MF exposure levels were assessed in the various factories and workshops of shielded metal arc welding in Ghana (Sawyerr et al. 2021). The magnetic flux densities observed ranged from 4.01 ± 0.72 μT to 196.46 ± 4.86 μT. The estimated induced current densities in the welders' heads and trunks ranged from 0.01 to 0.62 mA/m^2 and 0.03 to 1.23 mA/m^2, respectively. These findings were within the ICNIRP's reference level and basic restriction guidelines. However, responses to the administered questionnaire indicated that radiological safety practices among welders were insufficient, with many reporting common symptoms associated with ELF-MF exposure. Raphela et al. (2013) evaluated occupational exposure to ELF-MFs among 88 randomly selected welders and fitters in a South African welding industry. Measurements were taken in three workstations and an office. The findings showed that magnetic field exposure levels were significantly higher in the welding workshops compared to the office, with welders and fitters exposed to an average of 7.6 μT in the workshops. Median electric fields were relatively low across all three workstations: −15.50 and −13.50 volts per metre (V/m) in two workshops, and 1.80 V/m in the office. Rathebe and Mbonane (2018) evaluated the emissions of ELF-MFs in 132 kV distribution substations within the Free State, South Africa. Four 132 kV substations were randomly selected across three prominent areas within the province. Each substation was divided into 50 (1 m^2) blocks from the floor plans, resulting in a total of 200 measurements collected using a Triefield meter model XE 100. The statistical analysis revealed a non-significant difference in ELF-MF emissions across the substations ($p < 0.39$). Area one exhibited a slightly higher mean emission value (1.26 μT) compared to areas two (1.15 μT) and three (1.14 μT). The exposure levels measured were below the ICNIRP guidelines for both workers and the general public.

3.4 FUTURE STUDIES

Future studies investigating the exposure and health effects of NIR in Africa should prioritize a comprehensive approach that addresses several key areas to ensure thorough and impactful research. These areas include expanding geographic coverage, enhancing methodological rigour, considering vulnerable populations, integrating multidisciplinary approaches, and fostering policy and regulatory development. It is crucial to expand the geographic coverage of studies on NIR exposure in Africa. Most existing research tends to be concentrated in a few urban areas or specific regions, leaving vast parts of the continent underrepresented. Future studies should aim to include rural areas, diverse climatic zones, and different socio-economic settings.

This broader geographic scope will provide a more representative understanding of exposure levels across the continent and help identify regional variations and hotspots that may require targeted interventions. Enhancing the methodological rigour of studies is essential for producing reliable and comparable data. Future research should standardize measurement techniques, duration, and equipment used for assessing NIR. This includes employing advanced technologies and validated instruments for accurate measurement of exposure levels. Additionally, longitudinal studies that track exposure over extended periods will be valuable in understanding the long-term health effects of NIR, as opposed to short-term cross-sectional studies that provide only a snapshot of exposure.

It is important to focus on vulnerable populations, such as children, pregnant women, and individuals with pre-existing health conditions. These groups may be more susceptible to the potential health effects of NIR. Studies should assess the exposure levels and health outcomes specifically in these populations to determine if they are at greater risk. Furthermore, investigating the differential impacts of NIR on these vulnerable groups can guide the development of targeted public health interventions and protective measures. Integrating multidisciplinary approaches will enrich the understanding of NIR exposure and its health effects. Collaboration between epidemiologists, engineers, physicists, public health experts, and social scientists can provide a holistic view of the issue. For instance, engineers can contribute to improving measurement tools and methods, while social scientists can explore public awareness and perceptions of NIR risks. This multidisciplinary approach can lead to more comprehensive risk assessments and innovative solutions to mitigate exposure. Fostering policy and regulatory development is critical to addressing NIR exposure and health effects in Africa. Future studies should not only focus on generating scientific evidence but also on translating research findings into actionable policies. This includes advocating for stricter regulations on radiation emissions, establishing exposure limits based on local context, and promoting public awareness campaigns. Engaging policymakers and stakeholders from the outset can ensure that research findings are effectively integrated into national health strategies and regulatory frameworks, ultimately protecting public health. By addressing these key areas, future studies on NIR exposure and health effects in Africa can provide a robust foundation for informed decision-making, effective public health interventions, and comprehensive regulatory policies that safeguard the well-being of all populations across the continent.

3.5 CONCLUSION

This chapter has provided a comprehensive overview of studies assessing NIR exposure and its potential health effects in various regions across Africa. The findings highlight significant variability in exposure levels, influenced by geographic, environmental, and infrastructural factors. Despite generally low exposure levels, certain hotspots with higher radiation intensities were identified, necessitating targeted interventions and continuous monitoring. This work underscores the importance of adopting standardized measurement techniques and conducting longitudinal studies to capture temporal variations and long-term health impacts.

Furthermore, the chapter emphasizes the need for focused investigations on vulnerable populations, integrating multidisciplinary approaches to enrich the understanding of NIR effects. The development and enforcement of robust regulatory frameworks, guided by empirical evidence, are crucial for mitigating risks and protecting public health. Overall, this chapter lays the groundwork for future studies and provides essential insights for policymakers, health practitioners, and researchers to address the challenges posed by NIR in Africa.

REFERENCES

Ajibare, A.T., Oladejo, S.O., Ekwe, S.O., Akinyemi, L.A. and Ramotsoela, D., 2023. Radiofrequency electromagnetic radiation exposure assessment, analysis, computation, and minimization technique in 5G networks: a perspective on QoS trade-offs. *SAIEE Africa Research Journal*, 114(4), pp.114–127.

Annida, D.B., Oladoja, S.A., Oni, O.M., Isreal, E.L., Ajayi, O.B. and Aliagan, A.O., 2021, October. Assessment of radiofrequency radiation exposure level from selected base transceiver stations in Ogbomoso, Oyo State, Nigeria. In *Journal of Physics: Conference Series* (Vol. 2034, No. 1, p. 012014). IOP Publishing.

Aydin, E. and Cagiltay, N., 2012. A new RF and microwave engineering course enriched with advanced technologies. *Computer Applications in Engineering Education*, 20(4), pp.634–645.

Badru, R., Akinwale, O., Salau, A., Olorunyomi, K., Alwadood, J. and Atijosan, A., 2019. Assessment of geo-spatial proximity and magnetic pollution from 132kV and 330kV power transmission lines to infrastructures in Osogbo, Nigeria. *Eskişehir Teknik Üniversitesi Bilim ve Teknoloji Dergisi B-Teorik Bilimler*, 7(1), pp.81–93.

Banik, S., Bandyopadhyay, S. and Ganguly, S., 2003. Bioeffects of microwave—a brief review. *Bioresource Technology*, 87(2), pp.155–159.

Chege, B.W., 2015. Analysis of radiation levels in Nairobi's central business district and the industrial area, Kenya (Doctoral dissertation, JKUAT).

Cho, J.H., Lee, J.E. and Ra, C.S., 2009. Microwave irradiation as a way to reutilize the recovered struvite slurry and to enhance system performance. *Journal of Animal Science and Technology*, 51(4), pp.337–342.

Deatanyah, P., Amoako, J.K., Amoatey, E., Osei, S., Azah, C.K., Dogbey, R., Fuseini, A.R., Quarshie, E. and Owusu-Banahene, J., 2020. Assessment of radiofrequency radiation within the vicinity of some GSM base stations in Ghana: a follow-up. *Radiation Protection Dosimetry*, 192(4), pp.413–420.

Foster, K.R. and Repacholi, M.H., 2004. Biological effects of radiofrequency fields: does modulation matter? *Radiation Research*, 162(2), pp.219–225.

Gallagher, R.P., Lee, T.K., Bajdik, C.D. and Borugian, M., 2010. Rayonnement ultraviolet. *Maladies chroniques au Canada*, 29(1), p.60.

Garba, N.N., Umar, S. and Zakari, Y.I., 2017. Assessment of radio-frequency radiation exposure from selected mobile base stations in Kaduna state, Nigeria. *Nigerian Journal of Scientific Research*, 16(2), pp.184–186.

Guan, W., Huang, L., Wen, S., Yan, Z., Liang, W., Yang, C. and Liu, Z., 2021. Robot localization and navigation using visible light positioning and SLAM fusion. *Journal of Lightwave Technology*, 39(22), pp.7040–7051.

Gupta, S., Sharma, R.S. and Singh, R., 2022. Non-ionizing radiation as possible carcinogen. *International Journal of Environmental Health Research*, 32(4), pp.916–940.

ICNIRP., 2020. Guidelines for limiting exposure to electromagnetic fields (100 kHz to 300GHz). *Health Physics*, 118, pp.483–524.

INTERPHONE Study Group., 2010. Brain tumour risk in relation to mobile telephone use: results of the INTERPHONE international case–control study. *International Journal of Epidemiology*, 39(3), pp.675–694.

Inyang, E.P., Peter, E. and William, P., 2017. Assessment of the exposure of radio frequency radiation from Wi-Fi routers in calabar metropolis, Nigeria. *Nigerian Journal of Scientific Research*, 16(4), pp.490–494.

Joyner, K.H., Van Wyk, M.J. and Rowley, J.T., 2014. National surveys of radiofrequency field strengths from radio base stations in Africa. *Radiation Protection Dosimetry*, 158(3), pp.251–262.

Kheifets, L. and Oksuzyan, S., 2008. Exposure assessment and other challenges in non-ionizing radiation studies of childhood leukaemia. *Radiation Protection Dosimetry*, 132(2), pp.139–147.

Kheifets, L., Repacholi, M., Saunders, R. and Van Deventer, E., 2005. The sensitivity of children to electromagnetic fields. *Pediatrics*, 116(2), pp.e303–e313.

Maduka, N.C., Abdulmuniru, I. and Ahmad, M.U., 2019. Assessment of human exposure to non-ionizing radiation emitted from mobile base Station in Gusua, Zamfara State. *International Journal of Innovative Science Engineering and Technology*, 6(8), pp.98–109.

Mohammed, M.O., Elzaki, A.A., Babiker, B.A. and Eid, O.I., 2022. Spatial variability of outdoor exposure to radiofrequency radiation from mobile phone base stations, in Khartoum, Sudan. *Environmental Science and Pollution Research*, 29, pp. 15026–15039.

Nahuku, A., Tembo, K., Katengeza, E. and Ngwira, F., 2018. Analysis of radiation levels from mobile phone base stations in selected urban areas of Blantyre. *Advances in Ecological and Environmental Research*, pp.145–154.

Ndoma, E., George, N., Nathaniel, E., Orosun, M., Agbo, E. and Offorson, G., 2024. Technological civilization and health impact assessment of non-ionizing radiation in Nigeria. *Polytechnica*, 7(1), p.3.

Nwamaka, I.F., Adeyemi, O.J. and Rasaq, E.K., 2023. Assessment of radiofrequency radiation level in the near field of selected mobile phones in Katsina State, Nigeria. *IMCC Journal of Science*, 3, pp.16–30.

Ojuh, O.D. and Isabona, J., 2015. Radio frequency EMF exposure due to GSM Mobile phones base stations: measurements and analysis in Nigerian environment. *Nigerian Journal of Technology*, 34(4), pp.809–814.

Olorunsola, A.B., Ikumapayi, O.M., Oladapo, B.I., Alimi, A.O. and Adeoye, A.O., 2021. Temporal variation of exposure from radio-frequency electromagnetic fields around mobile communication base stations. *Scientific African*, 12, p.e00724.

Otitoloju, A.A., Obe, I.A., Adewale, O.A., Otubanjo, O.A. and Osunkalu, V.O., 2010. Preliminary study on the induction of sperm head abnormalities in mice, Mus musculus, exposed to radiofrequency radiations from global system for mobile communication base stations. *Bulletin of Environmental Contamination and Toxicology*, 84, pp.51–54.

Oyewopo, A.O., Olaniyi, S.K., Oyewopo, C.I. and Jimoh, A.T., 2017. Radiofrequency electromagnetic radiation from cell phone causes defective testicular function in male Wistar rats. *Andrologia*, 49(10), p.e12772.

Pourang, A., Tisack, A., Ezekwe, N., Torres, A.E., Kohli, I., Hamzavi, I.H. and Lim, H.W., 2022. Effects of visible light on mechanisms of skin photoaging. *Photodermatology, Photoimmunology & Photomedicine*, 38(3), pp.191–196.

Raphela, S.F., Weyers, C. and Shale, K., 2013. Assessment of occupational exposure to electromagnetic fields in the welding industry. *Occupational Health Southern Africa*, 19(1), pp.13–19.

Rathebe, P.C. and Mbonane, T.P., 2018, November. Emission levels of ELF magnetic fields from 132 kV distribution substations. In 2018 IEEE 4th Global Electromagnetic Compatibility Conference (GEMCCON) (pp. 1–3). IEEE.

Rathebe, P.C., 2020, September. Emission levels of ELF magnetic fields under medium voltage power-lines in Ngodini, Mpumalanga Province. In 2020 International Symposium on Electromagnetic Compatibility-EMC EUROPE (pp. 1–4). IEEE.

Rathebe, P.C., 2022. Characterization of radiofrequency electromagnetic fields in two unoccupied residential buildings proximal to the network base-station. In 2022 International Conference on Electrical, Computer and Energy Technologies (ICECET) (pp. 1–4). IEEE.

Rathebe, P.C., 2023, September. Exposure levels of radiofrequency electromagnetic fields from mobile base stations in Mpumalanga province, South Africa. In Journal of Physics: Conference Series (Vol. 2591, No. 1, p. 012035). IOP Publishing.

Sawyerr, A., Amoako, J.K., Fletcher, J.J., Sosu, E., Portuphy, M. and Kalognia, J., 2021. Evaluation of occupational exposure to extremely low frequency magnetic fields in shield metal arc welding processing in Accra, Ghana. *Journal of Radiation and Nuclear Applications*, 6, pp.193–199.

Schieke, S.M., Schroeder, P. and Krutmann, J., 2003. Cutaneous effects of infrared radiation: from clinical observations to molecular response mechanisms. *Photodermatology, Photoimmunology & Photomedicine*, 19(5), pp.228–234.

Sklar, L.R., Almutawa, F., Lim, H.W. and Hamzavi, I., 2013. Effects of ultraviolet radiation, visible light, and infrared radiation on erythema and pigmentation: a review. *Photochemical & Photobiological Sciences*, 12, pp.54–64.

Thulu, F.G.D., Tembo, D., Nyirongo, R., Mzaza, P.J.C., Kamfosi, A. and Mawenda, U.C., 2023. Electromagnetic frequency pollution in Malawi: a case of electric field and magnetic flux density pollution in Southern Africa. *International Journal of Environmental Research and Public Health*, 20(5), p.4413.

World Health Organization, 2007. *International classification of functioning, disability, and health: children & youth version: ICF-CY.* Geneva: World Health Organization.

Zhi, W.J., Wang, L.F. and Hu, X.J., 2017. Recent advances in the effects of microwave radiation on brains. *Military Medical Research*, 4, pp.1–14.

4 Health effects of ionizing radiation among the general public and workers

4.1 INTRODUCTION OF HEALTH EFFECTS AMONG THE GENERAL PUBLIC AND WORKERS' EXPOSURE TO IONIZING RADIATION

The exposure to ionizing radiation (IR) poses significant health risks to both the general public and occupational workers. This is particularly relevant in Africa, where diverse sources of IR exist, ranging from natural environmental sources to human-made activities. Understanding the health implications of such exposure is critical for developing effective safety measures and regulatory frameworks. This chapter explores the health effects of IR among the general public and workers in Africa, emphasizing the need for comprehensive risk assessment and robust policy implementation.

IR includes a variety of forms such as alpha particles, beta particles, gamma rays, and X-rays, all of which possess enough energy to ionize atoms and molecules, potentially leading to cellular and DNA damage (Chaudhary and Kumar 2023). The primary sources of IR in Africa include natural background radiation, medical diagnostic and therapeutic procedures, industrial applications, and, to a lesser extent, nuclear facilities (Einstein 2023). Natural background radiation predominantly arises from cosmic rays, terrestrial radionuclides, and radon gas, which is particularly prevalent in certain African regions due to geological formations (Aguko 2024). Medical applications of IR are increasingly common across the continent, with diagnostic radiology and radiotherapy being the primary contributors (Grover et al. 2015). While these technologies are indispensable for modern healthcare, they also pose significant risks to both patients and healthcare workers if not properly managed. The overuse or inappropriate use of diagnostic imaging and therapeutic radiation can lead to deterministic effects such as skin burns and acute radiation syndrome, as well as stochastic effects including an increased risk of cancer (Mohan and Chopra 2022). The need for stringent safety protocols and regular training for medical professionals cannot be overstated (Le Heron et al. 2010).

In the industrial setting, IR is utilized in various applications including radiography, nuclear power generation, and mining operations (Gupta 2013). Workers in

DOI: 10.1201/9781003508588-4

these industries are at a heightened risk of exposure, often encountering higher levels of radiation than the general public. Studies have shown that prolonged exposure to IR can lead to chronic health conditions such as cancer, cardiovascular diseases, and cataracts (Baselet et al. 2016; Little et al. 2021; Jahng et al. 2024). Moreover, the lack of adequate protective measures and monitoring in some African countries exacerbates these risks, highlighting the necessity for improved occupational health and safety standards, and regular health surveillance for workers. Despite the recognized hazards, there is a concerning lack of comprehensive data on radiation exposure and its health effects in many African countries. Existing studies are often fragmented and limited in scope, impeding the development of evidence-based policies and interventions. For instance, the monitoring of environmental radiation levels and occupational exposure is not uniformly conducted across the continent, leading to gaps in risk assessment and management (Makhijani et al. 2006). This suggests the importance of establishing robust radiation protection infrastructure, including standardized measurement techniques, centralized databases, and coordinated research efforts. Furthermore, public awareness of the risks associated with IR remains low in many African communities. Misconceptions and lack of knowledge about radiation safety can lead to unsafe practices and reluctance to adhere to recommended guidelines. Education and outreach programmes are crucial for enhancing public understanding of radiation risks and promoting protective behaviours. Engaging community leaders and leveraging media platforms can effectively disseminate information and foster a culture of safety (Dad et al. 2017).

4.2 OVERVIEW OF HEALTH EFFECTS AMONG THE GENERAL PUBLIC AND WORKERS' EXPOSURE TO IONIZING RADIATION SOURCES

Acute exposure to high levels of IR can result in immediate health effects, often referred to as acute radiation syndrome (ARS). ARS occurs when an individual is exposed to a large dose of radiation over a short period. Symptoms can range from mild (nausea and vomiting) to severe (haematopoietic, gastrointestinal, and neurovascular syndromes), depending on the radiation dose (Hu et al. 2020). For instance, the Chernobyl nuclear disaster provided a tragic example, where those in close proximity to the explosion experienced severe ARS, with many succumbing to the effects shortly after exposure (Higginbotham 2019). Long-term health effects from IR are primarily associated with the development of cancers. Epidemiological studies, such as those conducted on atomic bomb survivors in Hiroshima and Nagasaki, have consistently shown an increased incidence of various cancers, including leukaemia, breast, thyroid, and lung cancers, decades after exposure (Douple et al. 2011; Ozasa 2016). The linear no-threshold (LNT) model suggests that even low doses of radiation can increase cancer risk proportionally, with no safe threshold (Tharmalingam et al. 2019).

In Africa, uranium mining regions have reported elevated levels of IR in the environment, raising concerns about long-term cancer risks among local populations (Winde et al. 2017). The scarcity of robust epidemiological data in many African countries, however, complicates the assessment of these risks, necessitating more extensive research and monitoring. In addition to cancer, IR exposure is associated with non-cancer health effects. Cardiovascular diseases, particularly heart diseases and strokes, have been linked to radiation exposure. Studies on Hiroshima and Nagasaki survivors revealed a significant increase in the risk of cardiovascular diseases among those exposed to radiation (Douple et al. 2011). Other studies have shown that even moderate doses of radiation can lead to increased mortality from cardiovascular diseases (Little et al. 2021). IR can also cause genetic mutations and teratogenic effects, affecting not only the individuals exposed but also their offspring. Animal studies have demonstrated that exposure to radiation can lead to genetic mutations passed on to subsequent generations (Asimov and Dobzhansky 2023). In humans, there is evidence suggesting that parental exposure to radiation may result in congenital malformations and an increased risk of cancer in offspring (Little et al. 2013). This highlights the need for protective measures, particularly for pregnant women and individuals of reproductive age.

The psychological impact of radiation exposure, while less studied, is significant. Fear of radiation-related health effects can lead to chronic stress and anxiety. This was evident in populations affected by the Fukushima nuclear disaster, where increased rates of depression and post-traumatic stress disorder (PTSD) were reported among evacuees (Orui et al. 2021). The social stigma associated with radiation exposure can also lead to social isolation and discrimination, exacerbating mental health issues. Comparative studies have shown that the health effects of IR exposure are consistent across different populations and exposure scenarios. For example, studies on radiation workers, medical patients undergoing radiotherapy, and populations living near nuclear facilities all report similar patterns of increased cancer risk and other health effects (Choi et al. 2015; Vaiserman et al. 2018). These findings reinforce the universality of IR's health risks and the need for harmonized global standards in radiation protection. In Africa, research is still emerging, but studies indicate that the region faces unique challenges due to limited infrastructure for radiation monitoring and public health surveillance. Initiatives like the African Radiation Protection Association (ARPA) aim to address these gaps by promoting radiation protection and safety standards across the continent (Coraddu et al. 2015). Addressing the unique challenges through research, monitoring, and international collaboration is crucial to safeguarding public health against the effects of IR in Africa.

Globally, various occupations experience exposure from low to higher radiation dose. Healthcare workers who experienced accidental exposure during radiation therapy procedures have reported acute symptoms, underscoring the immediate dangers of high-dose exposure (Dörr and Meineke 2011). Also, long-term exposure to lower doses of IR has been extensively studied, revealing a significant increase in cancer risk among exposed workers. Epidemiological studies on nuclear industry workers and medical professionals consistently show elevated incidences of leukaemia, thyroid, lung, and breast cancers compared to the general population (Gillies and Haylock 2014; Boice et al. 2022). For instance, a cohort study of over 200,000 nuclear workers across three continents demonstrated a clear dose-response relationship between radiation exposure and cancer mortality (Cardis et al. 1995; Brüske-Hohlfeld 2009). Similarly, healthcare workers who perform frequent fluoroscopic procedures are at an increased risk of developing radiation-induced cancers (Ko et al. 2018). Protective measures, such as lead aprons and thyroid shields, are essential but not always sufficient to eliminate the risk, highlighting the need for improved safety standards and protocols (Cheon et al. 2018).

Occupational IR exposure has been linked to cardiovascular diseases. Research indicates that even moderate doses can increase the risk of heart disease and stroke. The Life Span Study of atomic bomb survivors revealed that cardiovascular disease risk increases with dose, with significant impacts seen at doses as low as 0.5 Gy (Little et al. 2022). Nuclear industry workers and radiologists, who receive cumulative radiation doses over their careers, may therefore be at heightened risk of cardiovascular conditions (Yoshinaga et al. 2004; Manenti et al. 2024). Moreover, occupational exposure IR can lead to cataracts, a non-malignant yet serious health effect. Studies have shown that radiation doses as low as 0.5 Gy can increase the risk of lens opacities, and workers exposed to radiation during interventional radiology procedures are particularly vulnerable (Azizova et al. 2018; Little et al. 2018). Research has demonstrated that radiation exposure can cause DNA damage, which may lead to genetic mutations in the exposed individual's offspring (Kamiya et al. 2015). Female radiologists and nuclear workers are at risk of radiation-induced infertility and adverse pregnancy outcomes, such as miscarriages and congenital abnormalities in their children (Latini et al. 2012). Studies across different occupational settings consistently report similar health effects associated with IR exposure. The International Agency for Research on Cancer (IARC) conducted a comprehensive review, confirming increased cancer risks among radiation-exposed workers in various industries, including healthcare, nuclear energy, and aviation (Zeeb et al. 2023). Additionally, comparative studies indicate that while regulatory frameworks vary globally, the health risks remain consistent (Vaiserman et al. 2018).

4.3 SOURCES SUCH AS NUCLEAR POWER PLANTS, UV RADIATION, X-RAY MACHINES, GOLD MINING, ETC.

In Africa, various sources contribute to IR exposure, including nuclear power plants, ultraviolet (UV) radiation, X-ray machines, and mining activities, particularly gold mining. This section explores these sources, highlighting their impacts, regulatory frameworks, and comparing findings with similar studies globally.

4.3.1 Nuclear power plants

Nuclear power plants are a significant source of IR. Although Africa has a limited number of nuclear power plants compared to other continents, countries like South Africa have operational nuclear facilities such as the Koeberg Nuclear Power Station. These plants produce radiation primarily through the fission of uranium-235, generating neutrons, gamma rays, and beta particles (Karmaker et al. 2021). Nuclear power plants are designed with extensive safety measures to contain radiation and prevent exposure to workers and the public. However, accidental releases can occur, as evidenced by the Fukushima and Chernobyl disasters. In Africa, continuous monitoring and adherence to international safety standards set by the International Atomic Energy Agency (IAEA) are crucial to mitigate risks. A study by Lloyd et al. (2013) highlighted that routine emissions from nuclear power plants are generally low and within safe limits. However, the potential for catastrophic accidents necessitates rigorous safety protocols and emergency preparedness. South Africa, being the leading nuclear power producer in Africa, has robust regulatory frameworks and continuous monitoring to ensure public safety (Mangena 2021).

4.3.2 Ultraviolet (UV) radiation

UV radiation is a form of electromagnetic radiation from the sun and artificial sources like tanning beds. While primarily non-ionizing, UV radiation at the shorter wavelengths (UV-C and some UV-B) possesses enough energy to cause ionization and damage biological tissues, leading to skin cancer and other health issues (Chinnasamy et al. 2022). Africa's geographical location near the equator results in high UV radiation levels, increasing the risk of skin cancer among the population. The World Health Organization (WHO) reports a higher incidence of skin cancer in regions with intense UV exposure. Preventive measures such as public education on UV protection, use of sunscreen, and protective clothing are essential to reduce the risk (Sultana 2020). Comparative studies indicate that while UV radiation is a global concern, regions with higher solar intensity, such as Africa, face more significant health challenges. In countries like Australia, public health campaigns have successfully raised awareness and reduced skin cancer rates, offering a model for similar initiatives in Africa (Miligi 2020).

4.3.3 X-ray machines

X-ray machines are widely used in medical diagnostics and industrial applications. They produce IR by accelerating electrons towards a metal target, generating X-rays through the sudden deceleration of electrons. In Africa, the increasing demand for medical imaging has led to the proliferation of X-ray machines in hospitals and clinics (Abdul-Razak 2019). The use of X-ray machines, while beneficial, poses risks to both patients and healthcare workers. Prolonged or repeated exposure to X-rays can increase the risk of cancers and other radiation-induced diseases (Wall et al. 2006). Regulatory bodies like the International Commission on Radiological Protection (ICRP) provide guidelines on safe exposure levels and the use of protective measures to minimize risks. A study conducted in Ghana by Botwe et al. (2015) revealed that adherence to radiation protection protocols varied significantly across healthcare facilities, with some exceeding recommended dose limits. This highlights the need for standardized training and stricter enforcement of safety regulations to protect both patients and healthcare workers.

4.3.4 Gold mining

Gold mining is a significant economic activity in several African countries, including South Africa, Ghana, and Tanzania. This activity often involves the extraction of gold from ore, which contains naturally occurring radioactive materials

(NORMs) such as uranium and thorium. The mining and processing of gold release these radionuclides into the environment, posing health risks to miners and nearby communities (Kamunda 2017). Exposure to IR from NORMs in gold mining can lead to lung cancer, especially among workers who inhale dust containing radioactive particles. A study by Lindsay et al. (2022) in South Africa's gold mines indicated elevated levels of radon, a radioactive gas, in underground mining areas, increasing cancer risks among miners. Risks associated with NORMs in mining are not unique to Africa. Similar issues have been reported in uranium mining regions in Australia and Canada, where stringent regulations and continuous monitoring have been implemented to protect workers (Semenova et al. 2020; Chen 2023). Effective regulation and monitoring are critical in managing IR exposure from these sources. In Africa, regulatory bodies such as the South African National Nuclear Regulator (NNR) and Ghana Atomic Energy Commission (GAEC) oversee the safe use of radiation in various sectors. These organizations adhere to international standards set by the IAEA and ICRP, implementing guidelines to protect workers and the public.

4.4 REVIEW OF STUDIES ON LONG-TERM EFFECTS AMONG THE GENERAL PUBLIC AND WORKERS' EXPOSURE TO IONIZING RADIATION

One of the most significant long-term health effects of IR exposure is an increased risk of cancer. The International Agency for Research on Cancer (IARC) classifies IR as a Group 1 carcinogen, meaning there is sufficient evidence of its carcinogenicity in humans. Studies have shown that exposure to low doses of radiation, such as those from medical diagnostic procedures, can lead to an increased risk of developing cancers such as leukaemia, thyroid cancer, breast cancer, and lung cancer (Linet et al. 2012; Ali et al. 2020). In Africa, the lack of adequate radiation protection measures and regulatory frameworks exacerbates this risk, particularly in regions with high levels of natural background radiation, such as parts of Nigeria and South Africa (Yusuf et al. 2020). In addition to cancer, long-term exposure to IR has been associated with an increased risk of cardiovascular diseases. Epidemiological studies have indicated that even low to moderate doses of radiation can damage the cardiovascular system, leading to conditions such as coronary artery disease and stroke (Little et al. 2015). In Africa, the public health burden of cardiovascular diseases is already high, and radiation exposure could further exacerbate this issue. Studies specific to the African context are limited, but data from other regions suggest a need for cautious monitoring and preventive measures.

IR can also cause genetic damage, leading to mutations that may be passed on to future generations. While the evidence for heritable effects in humans is less conclusive than for cancer, studies on atomic bomb survivors and individuals exposed to radiation in occupational settings have shown an increased frequency of genetic mutations (Creager 2014). In Africa, where genetic disorders already pose a significant health challenge, radiation-induced genetic damage could have profound implications for public health.

Workers in industries involving radiation, such as healthcare, mining, and nuclear power, are at a heightened risk of developing radiation-induced cancers. Studies among medical radiologists and radiologic technologists have consistently shown elevated risks of cancers, particularly leukaemia and thyroid cancer (Yoshinaga et al. 2004). In Africa, the increasing deployment of radiological equipment in healthcare settings necessitates stringent occupational safety standards to protect workers. Unfortunately, many African countries lack comprehensive radiation protection programmes, leaving workers vulnerable to long-term health effects. Long-term exposure to IR among workers can also lead to non-cancerous health effects, including cataracts, skin injuries, and chronic radiation dermatitis. For instance, interventional cardiologists who frequently perform fluoroscopy-guided procedures have been shown to have a higher incidence of cataracts (Rajabi et al. 2015). In the mining sector, workers exposed to radon gas, a natural radioactive decay product, are at risk of developing lung diseases, including pneumoconiosis and chronic obstructive pulmonary disease (COPD) (Rahman et al. 2001). These occupational health risks necessitate robust monitoring and preventive strategies to mitigate long-term impacts.

Workers exposed to IR may also suffer from chronic conditions such as radiation-induced fibrosis, cardiovascular diseases, and thyroid dysfunctions (Patel et al. 2015). These conditions often develop years after initial exposure, underscoring the need for long-term health surveillance and follow-up. In Africa, where healthcare infrastructure may be limited, ensuring continuous monitoring and healthcare support for radiation-exposed workers is a significant challenge.

4.5 REVIEW OF STUDIES ON SHORT-TERM EFFECTS AMONG THE GENERAL PUBLIC AND WORKERS' EXPOSURE TO IONIZING RADIATION

Medical procedures such as X-rays, computed tomography (CT) scans, and radiotherapy are common sources of IR exposure. Healthcare workers, such as radiologists, technologists, and nurses, are routinely exposed to scattered waves of radiation from diagnostic and therapeutic equipment (Chakole et al. 2022). Patients undergoing clinical imaging procedures also receive doses of IR, which can vary significantly depending on the type and frequency of the procedures. Apart from medical sources, Africa is rich in mineral resources, and mining activities, particularly gold and uranium mining, are prevalent. These activities expose workers to natural radionuclides present in ores and surrounding rocks. Miners can inhale radon gas, a decay product of uranium, which poses significant health risks (Semenova et al. 2020). Additionally, dust containing radioactive particles can lead to internal exposure when inhaled or ingested.

Industries such as oil and gas, construction, and manufacturing use radioactive materials for various purposes, including non-destructive testing and quality control. Workers in these industries are at risk of accidental exposure if proper safety protocols are not followed. Industrial radiography, for instance, uses gamma sources to inspect the integrity of materials, and improper handling of these sources can lead to acute radiation exposure (Reis et al. 2024). ARS is one of the most severe short-term effects of high-dose IR exposure. It occurs when the entire body or a significant portion is exposed to a high dose of radiation over a short period. Symptoms of ARS appear in stages and include nausea, vomiting, diarrhoea, and skin burns. Severe cases can lead to bone marrow failure, causing infections, bleeding, and even death (Dainiak and Sorba 1997). Localized radiation injuries occur when a specific body part receives a high dose of radiation. This can happen during medical procedures, such as interventional radiology, where high doses are targeted at a specific area. Symptoms include erythema (skin reddening), blistering, hair loss, and, in severe cases, tissue necrosis (Bennardo et al. 2021). Exposure to radon gas, particularly in mining environments and poorly ventilated homes, can cause respiratory issues. Radon decay products can be inhaled, leading to lung tissue damage and an increased risk of lung cancer. Short-term exposure to high levels of radon can result in respiratory symptoms such as coughing, wheezing, and shortness of breath (Mukharesh et al. 2022). Healthcare and industrial workers can suffer from radiation burns and skin damage due to accidental exposure to high doses of IR. This is particularly common in settings where radioactive sources are handled without adequate protection.

A study conducted in several hospitals in Sudan measured radiation doses received by healthcare workers in radiology departments. The results indicated that the mean monthly radiation doses were within the recommended limits set by the ICRP, but there was a need for improved protective measures and training to further reduce exposure (Suliman et al. 2019). In Tanzania's uranium mines, Jackson et al. (2022) highlighted the elevated risk of lung cancer among miners due to prolonged exposure to radon gas and radioactive dust. The study emphasized the importance of regular health screenings, improved ventilation, and the use of respiratory protection to mitigate these risks. The short-term effects of IR exposure among the general public and workers in Africa are significant and varied, depending on the source and level of exposure. ARS, localized radiation injuries, respiratory issues, and radiation burns are some of the immediate health effects observed within the limited Africa literature.

4.6 REVIEW OF STUDIES ON RISK PERCEPTION OF IONIZING RADIATION SOURCES AMONG THE GENERAL PUBLIC AND WORKERS

Cumulative exposure to IR leads to oxidative damage and genetic mutations, which are precursors to various diseases, including cancer (Tharmalingam et al. 2017). It is crucial to understand the concept of "IR," its harmful effects, and strategies for mitigation. The study by Ruth et al. (2020) focused on assessing the awareness of IR, its harmful impacts, and preventive measures among quarry workers in Nyamira County, Kenya. Using an analytical cross-sectional study design, data were gathered through questionnaires. The findings indicated that most quarry workers lacked adequate understanding of IR, its harmful effects, and preventive measures. However, education level and work experience were positively correlated with better understanding of IR, its harmful effects, and mitigation strategies, regardless of gender and age. This underscores the need for stakeholders to enhance awareness and education on IR, its effects, and preventive measures among quarry workers and the general public.

A study in the Cape Coast Ghana utilized a questionnaire-based approach, engaging 340 consenting participants from four hospitals to assess the perception of non-radiation healthcare workers about radiation (Edzie et al. 2021). A stratified sampling method determined the number of respondents from each hospital. Responses were collected via Google Forms through WhatsApp-administered questionnaires and systematically selected to achieve the sample size. Nurses comprised the largest group of respondents (n=166, 48.8%), followed by allied health professionals. The knowledge of participants about radiation issues was high, with over 65% demonstrating substantial awareness. The majority of non-radiation healthcare workers exhibited a high level of knowledge about radiation issues, and their perceptions were positive. A study in Nigeria evaluated the awareness of radon and its associated health risks among employees of Obafemi Awolowo University (OAU) in Ile-Ife. Participants included both academic and non-academic staff from the 13 faculties. A semi-structured questionnaire was administered to a sample of staff members based on the required sample size. Results showed that only 42% of respondents were aware of radon, and of these, 43.8% knew about the health risks associated with radon (Afolabi et al. 2015). In Zamfara state of Nigeria, Bartrem (2017) evaluated the perception and awareness of residents near mines regarding potential contamination and its adverse effects. Observational data were collected from April 2015 to January 2016, using two separately structured questionnaires. The target audience included inhabitants around mines and health personnel from stratified randomly sampled health centres that treat ailments related to mining activities. The study involved 217 respondents, with 87 health personnel from health centres and 130 local residents participating in the survey. The results, based on Likert scores, indicated that about 80% of respondents believed that mining in their areas could expose them to various radiological health impacts and environmental contamination. Health personnel responses revealed evidence of numerous ailments, including widespread Upper Respiratory Tract Infections (URTI) linked to aerosolized particulate radionuclides from mine tailings, as well as severe outcomes such as death.

In Khartoum Sudan, a cross-sectional study was conducted using a structured questionnaire, which included 13 questions covering demographic data, basic knowledge of radiation safety, methods of dose reduction, and attitudes towards radiation safety. A total of 167 dental practitioners participated in the study (Elmukhtar 2015). Half of the respondents were unaware that the thyroid gland is the most radiosensitive organ in the head and neck region. Forty-four percent believed that the dose for panoramic radiography was higher than for full-mouth periapical radiographs. Forty percent did not know about the relationship between the length of the X-ray cone and the radiation dose, and of the 44% who did, many could not explain that a long cone is more effective in reducing the dose. Forty-seven percent were unaware of the relationship between film speed and radiation dose, and of the 44% who were aware, many could not explain that fast film reduces patient dose more effectively. Sixty-six percent were unaware of the relationship between X-ray tube collimation and patient dose, and of the 46% who knew, many could not explain that rectangular collimation is more effective for dose reduction. Seventy-two percent did not know the safe distance from the radiation source.

4.7 FUTURE STUDIES

Understanding the risk perception of IR sources among the general public and workers in Africa is critical for developing effective radiation protection policies and health education programmes. As IR poses significant health risks, particularly in settings with limited regulatory oversight and awareness, future studies should focus on comprehensive assessments of knowledge, attitudes, and practices (KAP) related to radiation exposure. Conducting extensive KAP surveys among different demographic groups, including healthcare workers, miners, residents near nuclear facilities, and the general public, will provide insights into the level of awareness and misconceptions about IR. These surveys should cover various IR sources, including medical radiography, nuclear power plants, and naturally occurring radon gas.

Investigating how cultural beliefs and practices influence the perception of radiation risks in different African communities. Understanding cultural factors can help tailor public health messages and interventions to be more culturally sensitive and effective. Assessing the effectiveness of current educational and training programmes on radiation safety for both the general public and occupational groups. This includes evaluating the impact of formal education, workplace training, and public health campaigns on improving radiation risk perception and safety practices. Analysing the role of media in shaping public perceptions of IR. Future studies should also examine how different media platforms (e.g., television, radio, social media) disseminate information about radiation risks and how this information is received and interpreted by various audiences.

Evaluating the existing policy and regulatory frameworks governing radiation protection in African countries. Studies should assess the effectiveness of these frameworks in mitigating radiation risks and identify gaps that need to be addressed to enhance public and occupational safety. Utilizing both quantitative and qualitative research methods will provide a comprehensive understanding of radiation risk perception. Quantitative methods, such as surveys and questionnaires, can quantify levels of knowledge and attitudes, while qualitative methods, such as interviews and focus groups, can explore deeper insights and contextual factors influencing perceptions. Conducting longitudinal studies to track changes in risk perception over time, particularly before and after the implementation of educational or policy interventions. This will help in understanding the long-term impact of these interventions on radiation awareness and safety behaviours. Comparing risk perception and safety practices across different regions and countries within Africa. Such comparative studies can identify best practices and successful strategies that can be replicated in other settings.

Developing targeted public health campaigns that address the specific knowledge gaps and misconceptions identified through research. These campaigns should use culturally appropriate messages and leverage trusted communication channels to reach diverse audiences. Implementing comprehensive training programmes for workers in high-risk occupations, such as healthcare and mining, to ensure they are adequately informed about radiation risks and safety measures. Continuous professional development and certification programmes can help maintain high safety standards. Advocating for stronger regulatory frameworks and enforcement mechanisms to ensure compliance with radiation safety standards. Policymakers should be informed by research findings to develop evidence-based regulations that protect both workers and the general public. Encouraging community participation in radiation safety initiatives. Involving community leaders and local organizations can enhance the effectiveness of public health interventions and foster a culture of safety and awareness.

4.8 CONCLUSION

The health effects of IR among the general public and workers in Africa present a significant public health challenge that necessitates urgent attention and action. Exposure to IR, whether from natural sources like radon or human activities such as medical procedures and mining operations, poses serious risks, including cancer and other radiation-induced illnesses. Studies have consistently highlighted that awareness and understanding of these risks vary widely across different populations and occupational groups in Africa. This variability underscores the need for targeted education and training programmes to improve knowledge and promote radiation safety practices. Healthcare workers, miners, and other high-risk groups need to be particularly well-informed about the potential health impacts and the measures they can take to mitigate these risks.

Effective regulatory frameworks and enforcement mechanisms are critical for ensuring radiation safety. Governments and relevant authorities must prioritize the development and implementation of robust regulations that are informed by the latest scientific research. These regulations should be accompanied by regular monitoring and evaluation to ensure compliance and effectiveness. Public health initiatives should also focus on raising awareness among the general population. Campaigns that provide clear and accessible information about the sources of IR, the associated health risks, and protective measures can empower individuals to make informed decisions and take proactive steps to reduce their exposure. In the healthcare sector, efforts should be made to minimize unnecessary exposure to IR through the adoption of best practices and the use of advanced technologies that offer lower radiation doses. Similarly, in industries such as mining, stringent safety protocols and continuous monitoring are essential to protect workers from excessive radiation exposure.

Collaboration between international organizations, governments, researchers, and communities is vital to address the complex issues surrounding IR in Africa. By sharing knowledge, resources, and expertise, we can develop comprehensive strategies that effectively reduce the health risks associated with IR. Addressing the health effects of IR in Africa requires a multifaceted approach that includes education, regulation, and community engagement. By enhancing awareness, improving safety practices, and enforcing stringent regulations, we can protect the health and well-being of both the general public and workers exposed to IR. Ensuring a safer environment for all will ultimately contribute to the overall health and development of African nations.

REFERENCES

Abdul-Razak, W., 2019. Role extension opportunities in diagnostic radiography in Ghana and its potential to impact on patient care (Doctoral dissertation, Cardiff University).

Afolabi, O.T., Falode, D.T., Banjoko, B., Onayade, A.A. and Tobih, J.E., 2015. Radon risk awareness among university employees of Obafemi Awolowo university, Ile-Ife. American Association of Radon Scientist (AARST) Proceedings.

Aguko, W.O., 2024. Radiation exposure levels and excess lifetime cancer risk associated with soil, water and sub-surface rocks in Kargi Area, Marsabit-Kenya (Doctoral dissertation, JKUAT-COPAS).

Ali, Y.F., Cucinotta, F.A., Ning-Ang, L. and Zhou, G., 2020. Cancer risk of low dose ionizing radiation. *Frontiers in Physics*, 8, p.234.

Asimov, I. and Dobzhansky, T., 2023. *The genetic effects of radiation: exploring genetic implications of nuclear energy through speculative science fiction*. Washington DC: Good Press.

Azizova, T.V., Hamada, N., Grigoryeva, E.S. and Bragin, E.V., 2018. Risk of various types of cataracts in a cohort of Mayak workers following chronic occupational exposure to ionizing radiation. *European Journal of Epidemiology*, 33(12), pp.1193–1204.

Bartrem, C., 2017. Environmental heath and risk assessment: lead poisoning, remediation, and project sustainability in Zamfara, Nigeria and Lead-Arsenate Contamination in Washington, USA (Doctoral dissertation, University of Idaho).

Baselet, B., Rombouts, C., Benotmane, A.M., Baatout, S. and Aerts, A., 2016. Cardiovascular diseases related to ionizing radiation: the risk of low-dose exposure. *International Journal of Molecular Medicine*, 38(6), pp.1623–1641.

Bennardo, L., Passante, M., Cameli, N., Cristaudo, A., Patruno, C., Nisticò, S.P. and Silvestri, M., 2021. Skin manifestations after ionizing radiation exposure: a systematic review. *Bioengineering*, 8(11), p.153.

Boice Jr, J.D., Cohen, S.S., Mumma, M.T., Hagemeyer, D.A., Chen, H., Golden, A.P., Yoder, R.C. and Dauer, L.T., 2022. Mortality from leukemia, cancer and heart disease among US nuclear power plant workers, 1957–2011. *International Journal of Radiation Biology*, 98(4), pp.657–678.

Botwe, B.O., Antwi, W.K., Adesi, K.K., Anim-Sampong, S., Dennis, A.M., Sarkodie, B.D. and Opoku, S.Y., 2015. Personal radiation monitoring of occupationally exposed radiographers in the biggest tertiary referral hospital in Ghana. *Safety in Health*, 1, pp.1–7.

Brüske-Hohlfeld, I., 2009. Environmental and occupational risk factors for lung cancer. In: Verma, M. (eds), *Cancer Epidemiology. Methods in Molecular Biology*, vol 472. Humana Press. https://doi.org/10.1007/978-1-60327-492-0_1

Cardis, E., Gilbert, E.S., Carpenter, L., Howe, G., Kato, I., Armstrong, B.K., Beral, V., Cowper, G., Douglas, A., Fix, J. and Fry, S.A., 1995. Effects of low doses and low dose rates of external ionizing radiation: cancer mortality among nuclear industry workers in three countries. *Radiation Research*, 142(2), pp.117–132.

Chakole, M.V.S., Luharia, M.A., Tivaskar, M.S., Dhande, R. and Pathade, A., 2022. Radiation protection in diagnostic radiology. *Journal of Pharmaceutical Negative Results*, 13(8), pp.140–147.

Chaudhary, N. and Kumar, G., 2023. Mutagenic radiations: x-rays, ionizing particles, and ultraviolet. In T.A. Bhat and K.R. Hakeem (eds), *Biotechnologies and genetics in plant mutation breeding* (pp. 45–67). New York: Apple Academic Press.

Chen, J., 2023. A review of radon exposure in non-uranium mines—estimation of potential radon exposure in Canadian Mines. *Health Physics*, 124(4), pp.244–256.

Cheon, B.K., Kim, C.L., Kim, K.R., Kang, M.H., Lim, J.A., Woo, N.S., Rhee, K.Y., Kim, H.K. and Kim, J.H., 2018. Radiation safety: a focus on lead aprons and thyroid shields in interventional pain management. *The Korean Journal of Pain*, 31(4), pp.244–252.

Chinnasamy, S., Ramachandran, M. and Sivaji, C., 2022. A study on ultraviolet radiation and its effects. *REST Journal on Advances in Mechanical Engineering*, 1(2), pp.1–9.

Choi, T.A., Costes, S.V. and Abergel, R.J., 2015. *Understanding the health impacts and risks of exposure to radiation* (pp. 259–281). New York: Springer International Publishing.

Coraddu, M., Cottone, E., Levis, A., Lombardo, A., Marinelli, F. and Zucchetti, M., 2015. Electromagnetic fields (EMF) biological and health effects and the MUOS case. *Fresenius Environmental Bulletin*, 24(5a), pp.1896–1903.

Creager, A.N., 2014. Radiation, cancer, and mutation in the atomic age. *Historical Studies in the Natural Sciences*, 45(1), pp.14–48.

Dad, L., Royce, T.J., Morris, Z., Moran, M., Pawlicki, T., Khuntia, D., Hardenbergh, P., Cummings, B., Mayr, N. and Hu, K., 2017, April. Bridging innovation and outreach to overcome global gaps in radiation oncology through information and communication tools, trainee advancement, engaging industry, attention to ethical challenges, and political advocacy. In T.R. Williams and C.N. Coleman (eds), *Seminars in Radiation Oncology* (Vol. 27, No. 2, pp. 98–108). Philadelphia: WB Saunders.

Dainiak, N. and Sorba, S., 1997. Early identification of radiation accident victims for therapy of bone marrow failure. *Stem Cells*, 15(S1), pp.275–285.

Dörr, H. and Meineke, V., 2011. Acute radiation syndrome caused by accidental radiation exposure-therapeutic principles. *BMC Medicine*, 9, pp.1–6.

Douple, E.B., Mabuchi, K., Cullings, H.M., Preston, D.L., Kodama, K., Shimizu, Y., Fujiwara, S. and Shore, R.E., 2011. Long-term radiation-related health effects in a unique human population: lessons learned from the atomic bomb survivors of Hiroshima and Nagasaki. *Disaster Medicine and Public Health Preparedness*, 5(S1), pp.S122–S133.

Edzie, E.K.M., Dzefi-Tettey, K., Gorleku, P.N., Amankwa, A.T., Idun, E.A., Brakohiapa, E.K., Aidoo, E., Essoun, J.J., Quarshie, F., Kusodzi, H. and Asemah, A.R., 2021. Perception of non-radiation healthcare workers about radiation in Cape Coast Metropolis, Ghana. *Universal Journal of Public Health*, 9(2), pp.57–66.

Einstein, O.P., 2023. Cancer risks associated with exposure to background ionizing radiation in human habitat, soil, and food in Kenya (Doctoral dissertation, Kenyatta University).

Elmukhtar, Y.M., 2015. Knowledge and attitude of dentists regarding radiation safety in government dental clinics in Khartoum, Sudan (Dissertation, University of the Western Cape).

Gillies, M. and Haylock, R., 2014. The cancer mortality and incidence experience of workers at British Nuclear Fuels plc, 1946–2005. *Journal of Radiological Protection*, 34(3), p.595.

Grover, S., Xu, M.J., Yeager, A., Rosman, L., Groen, R.S., Chackungal, S., Rodin, D., Mangaali, M., Nurkic, S., Fernandes, A. and Lin, L.L., 2015. A systematic review of radiotherapy capacity in low-and middle-income countries. *Frontiers in Oncology*, 4, p.380.

Gupta, T., 2013. *Radiation, ionization, and detection in nuclear medicine* (pp. 451–494). Heidelberg: Springer.

Higginbotham, A., 2019. *Midnight in Chernobyl: the untold story of the world's greatest nuclear disaster*. New York: Random House.

Hu, S., Barzilla, J.E. and Semones, E., 2020. Acute radiation risk assessment and mitigation strategies in near future exploration spaceflights. *Life Sciences in Space Research*, 24, pp.25–33.

Jackson, M., Simon, C., Njale, E., Kanyanemu, A., Anatory, Y.S., Mhozya, M. and Semiono, P., 2022. Risk assessment of natural radionuclides and radon gas in the Artisanal and small-scale gold mine of Buhemba, Tanzania. *International Journal of Environmental Protection and Policy*, 10(3), pp.48–58.

Jahng, J.W., Little, M.P., No, H.J., Loo Jr, B.W. and Wu, J.C., 2024. Consequences of ionizing radiation exposure to the cardiovascular system. *Nature Reviews Cardiology*, pp.1–19.

Kamiya, K., Ozasa, K., Akiba, S., Niwa, O., Kodama, K., Takamura, N., Zaharieva, E.K., Kimura, Y. and Wakeford, R., 2015. Long-term effects of radiation exposure on health. *The Lancet*, 386(9992), pp.469–478.

Kamunda, C., 2017. Human health risk assessment of environmental radionuclides and heavy metals around a gold mining area in Gauteng province, south Africa (Doctoral dissertation, North-West University (South Africa) Mafikeng Campus).

Karmaker, N., Maraz, K.M., Islam, F., Haque, M.M., Razzak, M., Mollah, M.Z.I., Faruque, M.R.I. and Khan, R.A., 2021. Fundamental characteristics and application of radiation. *GSC Advanced Research and Reviews*, 7(1), pp.64–72.

Ko, S., Kang, S., Ha, M., Kim, J., Jun, J.K., Kong, K.A. and Lee, W.J., 2018. Health effects from occupational radiation exposure among fluoroscopy-guided interventional medical workers: a systematic review. *Journal of Vascular and Interventional Radiology*, 29(3), pp.353–366.

Latini, G., Dipaola, L., Mantovani, A. and Picano, E., 2012. Reproductive effects of low-to-moderate medical radiation exposure. *Current Medicinal Chemistry*, 19(36), pp.6171–6177.

Le Heron, J., Padovani, R., Smith, I. and Czarwinski, R., 2010. Radiation protection of medical staff. *European Journal of Radiology*, 76(1), pp.20–23.

Lindsay, R., Mngonyama, S., Molahlehi, P., Ngwadla, X.E. and Ramonnye, G.J., 2022. Pilot study of thoron concentration in an underground thorium mine. *Health Physics*, 123(4), pp.315–321.

Linet, M.S., Slovis, T.L., Miller, D.L., Kleinerman, R., Lee, C., Rajaraman, P. and Berrington de Gonzalez, A., 2012. Cancer risks associated with external radiation from diagnostic imaging procedures. *CA: A Cancer Journal for Clinicians*, 62(2), pp.75–100.

Little, M.P. and Lipshultz, S.E., 2015. Low dose radiation and circulatory diseases: a brief narrative review. *Cardio-Oncology*, 1, pp.1–10.

Little, M.P., Azizova, T.V. and Hamada, N., 2021. Low-and moderate-dose non-cancer effects of ionizing radiation in directly exposed individuals, especially circulatory and ocular diseases: a review of the epidemiology. *International Journal of Radiation Biology*, 97(6), pp.782–803.

Little, M.P., Goodhead, D.T., Bridges, B.A. and Bouffler, S.D., 2013. Evidence relevant to untargeted and transgenerational effects in the offspring of irradiated parents. *Mutation Research/Reviews in Mutation Research*, 753(1), pp.50–67.

Little, M.P., Kitahara, C.M., Cahoon, E.K., Bernier, M.O., Velazquez-Kronen, R., Doody, M.M., Borrego, D., Miller, J.S., Alexander, B.H., Simon, S.L. and Preston, D.L., 2018. Occupational radiation exposure and risk of cataract incidence in a cohort of US radiologic technologists. *European Journal of Epidemiology*, 33, pp.1179–1191.

Lloyd, R., 2013. An institutional approach to understanding the green paradox of nuclear power (Doctoral dissertation, Cardiff University).

Makhijani, A., Smith, B. and Thorne, M.C., 2006. *Science for the vulnerable: setting radiation and multiple exposure environmental health standards to protect those most at risk.* Institute for Energy and Environmental Research (Maryland).

Manenti, G., Coppeta, L., Kirev, I.V., Verno, G., Garaci, F., Magrini, A. and Floris, R., 2024, January. Low-dose occupational exposure to ionizing radiation and cardiovascular effects: a narrative review. *Healthcare*, 12(2), p.238. https://doi.org/10.3390/healthcare12020238

Mangena, J., 2021. Development of a strategic decision-making model for the South African nuclear sector (Doctoral dissertation, North-West University (South Africa)).

Miligi, L., 2020. *Ultraviolet radiation exposure: some observations and considerations, focusing on some Italian experiences, on cancer risk, and primary prevention. Environments*, 7(2), p.10. https://doi.org/10.3390/environments7020010

Mohan, S. and Chopra, V., 2022. Biological effects of radiation. In S. Dhoble, V. Chopra, V. Nayar, G. Kitis, D. Poelman, and H. Swart (eds), *Radiation dosimetry phosphors* (pp. 485–508). United Kingdom: Woodhead Publishing.

Mukharesh, L., Greco, K.F., Banzon, T., Koutrakis, P., Li, L., Hauptman, M., Phipatanakul, W. and Gaffin, J.M., 2022. Environmental radon and childhood asthma. *Pediatric Pulmonology*, 57(12), p.3165.

Orui, M., Nakayama, C., Moriyama, N., Tsubokura, M., Watanabe, K., Nakayama, T., Sugita, M. and Yasumura, S., 2021. Those who have continuing radiation anxiety show high psychological distress in cases of high post-traumatic stress: The Fukushima nuclear disaster. *International Journal of Environmental Research and Public Health*, 18(22), p.12048.

Ozasa, K., 2016. Epidemiological research on radiation-induced cancer in atomic bomb survivors. *Journal of Radiation Research*, 57(S1), pp.i112–i117.

Patel, Z., Huff, J., Saha, J., Wang, M., Blattnig, S., Wu, H. and Cucinotta, F., 2015. Evidence report: risk of cardiovascular disease and other degenerative tissue effects from radiation exposure (No. JSC-CN-34126).

Rahman, Q., Nettesheim, P., Smith, K.R., Seth, P.K. and Selkirk, J., 2001. International conference on environmental and occupational lung diseases. *Environmental Health Perspectives*, 109(4), pp.425–431.

Rajabi, A.B., Noohi, F., Hashemi, H., Haghjoo, M., Miraftab, M., Yaghoobi, N., Rastgou, F., Malek, H., Faghihi, H., Firouzabadi, H. and Asgari, S., 2015. Ionizing radiation-induced cataract in interventional cardiology staff. *Research in Cardiovascular Medicine*, 4(1), pp.1–6.

Reis, J.C.G.D., Pereira, J.C. and Dores, J.M.C.M.D., 2024. Comprehensive risk assessment of industrial radiography process within industry 5.0: towards continuous safety improvement. *Advances in Reliability, Safety and Security*, Part 8, pp.109–118.

Ruth, M.K., Warutere, P.N., Nyamari, J. and Arika, W., 2020. Demographic study of ionizing radiation exposure of quarry workers in some selected Quarries in Nyamira County, Kenya. *Environmental Health Insights*, 14, pp.1–15.

Semenova, Y., Pivina, L., Zhunussov, Y., Zhanaspayev, M., Chirumbolo, S., Muzdubayeva, Z. and Bjørklund, G., 2020. Radiation-related health hazards to uranium miners. *Environmental Science and Pollution Research*, 27, pp.34808–34822.

Suliman, I.I., Salih, L.H., Ali, D.M., Alaamer, A.S., Al-Rajhi, M.A., Alkhorayef, M. and Bradley, D.A., 2019. Occupational exposure in nuclear medicine and interventional cardiology departments in Sudan: are they following radiation protection standards? *Radiation Physics and Chemistry*, 160, pp.100–104.

Sultana, N., 2020. Sun awareness and sun protection practices. *Clinical, Cosmetic and Investigational Dermatology,* 13(1), pp.717–730.

Tharmalingam, S., Sreetharan, S., Brooks, A.L. and Boreham, D.R., 2019. Re-evaluation of the linear no-threshold (LNT) model using new paradigms and modern molecular studies. *Chemico-Biological Interactions*, 301, pp.54–67.

Tharmalingam, S., Sreetharan, S., Kulesza, A.V., Boreham, D.R. and Tai, T.C., 2017. Low-dose ionizing radiation exposure, oxidative stress and epigenetic programing of health and disease. *Radiation Research*, 188(4.2), pp.525–538.

Vaiserman, A., Koliada, A., Zabuga, O. and Socol, Y., 2018. Health impacts of low-dose ionizing radiation: current scientific debates and regulatory issues. *Dose-Response*, 16(3), p.1559325818796331.

Wall, B.F., Kendall, G.M., Edwards, A.A., Bouffler, S., Muirhead, C.R. and Meara, J.R., 2006. What are the risks from medical X-rays and other low dose radiation? *The British Journal of Radiology*, 79(940), pp.285–294.

Winde, F., Brugge, D., Nidecker, A. and Ruegg, U., 2017. Uranium from Africa–an overview on past and current mining activities: re-appraising associated risks and chances in a global context. *Journal of African Earth Sciences*, 129, pp.759–778.

Yoshinaga, S., Mabuchi, K., Sigurdson, A.J., Doody, M.M. and Ron, E., 2004. Cancer risks among radiologists and radiologic technologists: review of epidemiologic studies. *Radiology*, 233(2), pp.313–321.

Yusuf, S.D., Umar, I., Bukar, A. and Idris, M.M., 2020. Assessment of the knowledge and attitude to radiation safety standards of the radiological staff in Damaturu, Yobe State, Nigeria. *Asian Journal of Medical Principles and Clinical Practice*, 3(3), pp.1–7.

Zeeb, H., Merzenich, H., Blettner, M. and Wollschläger, D., 2023. Radiation epidemiology. In W. Ahrens and I. Pigeot (eds), *Handbook of epidemiology* (pp. 1–39). New York: Springer New York.

5 Health effects of non-ionizing radiation among the general public and workers

5.1 INTRODUCTION OF HEALTH EFFECTS AMONG THE GENERAL PUBLIC AND WORKERS' EXPOSURE TO NON-IONIZING RADIATION

Non-ionizing radiation (NIR) encompasses a broad spectrum of electromagnetic fields (EMF) that do not possess sufficient energy to ionize atoms or molecules (ICNIRP 2020). This spectrum includes ultraviolet (UV) radiation, visible light, infrared radiation, microwave radiation, radiofrequency (RF) radiation, and extremely low-frequency (ELF) fields. With the rapid technological advancement and urbanization in Africa, the exposure to various forms of NIR has significantly increased. This increase brings to the forefront concerns about potential health effects on both the general public and occupational workers. UV radiation from the sun is a major source of NIR. Africa, with its predominantly sunny climate, experiences high levels of UV exposure. The three types of UV radiation include UVA, UVB, and UVC, with UVB being the most biologically active and harmful to human skin (Lucas et al. 2006). RF radiation is commonly emitted from mobile phones, base stations, Wi-Fi routers, and other wireless communication devices. The widespread adoption of these technologies in Africa has led to increased RF exposure among the population (ICNIRP 2020). Microwaves, used in household appliances such as microwave ovens and in industrial applications, are another significant source of NIR. Occupational exposure is particularly relevant in industries that utilize high-frequency devices (WHO 2014). ELF fields are emitted from power lines, electrical wiring, and appliances. They are ubiquitous in modern living environments, contributing to continuous, low-level exposure (Ahlbom et al. 2008).

Excessive exposure to UV radiation can cause acute and chronic health effects. Acute effects include erythema (sunburn) and photokeratitis (inflammation of the cornea). Chronic exposure increases the risk of skin cancers, such as melanoma and non-melanoma skin cancers (NMSC) and accelerates skin ageing. UV radiation also contributes to ocular conditions like cataracts, which can lead to visual impairment (Roth et al. 2018). The health effects of RF radiation remain a topic of extensive research and debate. Acute exposure can lead to thermal effects, causing

DOI: 10.1201/9781003508588-5

tissue heating and potential burns. Non-thermal effects are more controversial, with studies suggesting potential links to headaches, sleep disturbances, cognitive impairment, and an increased risk of gliomas, a type of brain cancer (Khurana et al. 2009). Occupational exposure to microwave radiation can lead to thermal injuries, including burns and cataracts. Chronic exposure has been associated with potential neurological effects and reproductive health issues, though conclusive evidence is still lacking (WHO 2014). The International Agency for Research on Cancer (IARC) has classified RF radiation as possibly carcinogenic to humans (Group 2B) (IARC 2011). ELF fields have been studied for their potential association with childhood leukaemia and other health effects. While some epidemiological studies have indicated an increased risk of leukaemia among children exposed to high levels of ELF fields, the evidence is not definitive, and further research is needed to establish clear causal relationships (Kheifets et al. 2010).

Workers in various industries across Africa are at risk of NIR exposure. For example, healthcare workers using UV lamps for sterilization and RF devices for diagnostics are exposed to potentially harmful levels of NIR. Adequate safety measures, including protective equipment and exposure monitoring, are essential to minimize risks (ICNIRP 2020). Workers involved in the installation and maintenance of mobile phone towers and other telecommunication infrastructure are exposed to high levels of RF radiation. Training and adherence to safety protocols are crucial to prevent overexposure (Halgamuge et al. 2015). Employees in industries using microwave technology or operating near high-voltage power lines are exposed to microwave and ELF radiation, respectively (WHO 2014). The general public in Africa is increasingly exposed to NIR through various sources. The proliferation of mobile phones and wireless devices has led to widespread RF radiation exposure. Public awareness campaigns are needed to educate users on safe usage practices, such as limiting call duration and using hands-free devices (Ahlbom et al. 2008). Prolonged outdoor activities without adequate sun protection increase the risk of UV-induced skin damage. Public health initiatives should promote the use of sunscreen, protective clothing, and shaded areas to reduce UV exposure (Lucas et al. 2006). The use of household appliances such as microwave ovens and electrical devices contributes to exposure to microwave and ELF fields. Educating the public on safe appliance usage and maintaining safe distances can help minimize exposure (WHO 2014; Halgamuge et al. 2015).

5.2 OVERVIEW OF HEALTH EFFECTS AMONG THE GENERAL PUBLIC AND WORKERS' EXPOSURE TO NON-IONIZING RADIATION SOURCES

NIR is an essential concern in occupational and environmental health, particularly in Africa, where emerging technologies and infrastructure development have led to increased exposure. This discussion will explore the health effects of NIR among the general public and workers, focusing on sources such as RF radiation from mobile phones and base stations, and EMF. The context of African studies is explored, given the region's rapid industrialization and urbanization. NIR generally has longer wavelengths and lower frequencies. Common sources include radio waves, microwaves, infrared, UV light, and EMF. NIR is often categorized by frequency:

- ELF radiation from power lines and electrical appliances.
- RF radiation from mobile phones, Wi-Fi, and broadcasting stations.
- UV radiation from the sun and artificial sources such as tanning beds.

In Africa, the proliferation of mobile phones, Wi-Fi networks, and energy infrastructure development has heightened exposure to NIR for both workers and the general public. This increasing exposure raises concerns about the potential health risks posed by these technologies, which need to be explored in more depth.

Mobile phones are one of the most common sources of NIR exposure among the general population. RF radiation from mobile phones operates at frequencies ranging between 800 MHz and 2.6 GHz. Although mobile phones have revolutionized communication in Africa, concerns have emerged about the long-term health effects of their usage. Epidemiological studies have explored the potential link between mobile phone use and brain cancer, particularly glioma and acoustic neuroma. While no conclusive evidence has been found to establish a causal relationship, the IARC has classified RF radiation as "possibly carcinogenic to humans" (Group 2B). It is essential to continue monitoring long-term health effects through longitudinal studies. Exposure to EMFs from power lines, electrical appliances, and communication infrastructure is also widespread. Prolonged exposure to low-frequency EMFs has been suggested as a possible risk factor for conditions such as leukaemia, neurodegenerative disorders, and cognitive impairment, though evidence remains inconclusive.

In many African countries, informal settlements often exist near high-voltage power lines, raising concerns about the long-term health effects of sustained exposure to EMFs. Although data on the impact of EMF exposure in African populations is limited, studies from developed regions have suggested the possibility of adverse outcomes, including childhood leukaemia and sleep disturbances. There is an existing lack of occupational health policies and inadequate training on the safe handling of RF equipment among workers. This increases the risk of overexposure. Studies conducted in African countries, particularly South Africa, have pointed to a need for stricter regulation and protective measures for telecommunication workers (Rathebe 2023). The potential health effects of occupational exposure to RF radiation include headaches, fatigue, cognitive decline, and an increased risk of cancer, although definitive evidence remains limited.

Medical professionals who operate devices such as MRI machines, X-ray machines, and other diagnostic equipment are also exposed to NIR. While MRI machines use NIR in the form of RF waves, the magnetic fields generated during scans can pose health risks for workers regularly operating the equipment (Rathebe 2022a, b). Potential health impacts include thermal effects, dizziness, nausea, and sensory disturbances. In hospitals, throughout the entire continent, the demand for advanced diagnostic technology is growing, it is essential to provide medical workers with adequate training and protective equipment (Rathebe 2022). Many healthcare workers are unaware of the risks posed by prolonged exposure to NIR, which underscores the need for improved safety protocols and educational campaigns.

5.3 SOURCES SUCH AS ELECTRIC UTILITY STRUCTURES, TELECOMMUNICATION DEVICES, AND MRI

NIR from sources such as electric utility structures, telecommunication devices, and MRI machines has become increasingly prevalent in Africa due to technological advancement and infrastructure expansion. While these sources provide essential services, there are growing concerns about their potential health impacts on both the general public and workers, as well as the adequacy of regulations governing their safe use. Electric utility structures, such as power lines and transformers, emit ELF-EMF, which have been the subject of health concerns for decades. In Africa, the electrification of urban and rural areas is essential for economic development, yet it also leads to increased EMF exposure for both residents and utility workers. The expansion of electric utility infrastructure in Africa, especially in rapidly urbanizing areas, has increased public exposure to EMFs. Many informal settlements in African cities, such as Johannesburg and Lagos, are located near high-voltage power lines and transformers due to the haphazard nature of urban planning. This proximity raises concerns about long-term health effects. Although the evidence linking ELF-EMF exposure to cancer and other health conditions is inconclusive, studies in other regions have suggested a possible association with childhood leukaemia and neurodegenerative diseases (Kheifets et al. 2005). In South Africa, research has been conducted to understand the public's exposure to EMFs from electric utility structures. One study found that while EMF exposure levels in residential areas were generally below international safety limits, further research is needed to assess the long-term health risks, particularly for populations living in informal settlements (Dhlamini et al. 2015). With increasing electrification efforts across Africa, it is vital to conduct more local studies to determine whether existing exposure limits adequately protect public health.

Utility workers involved in the construction, maintenance, and repair of power lines are at a higher risk of occupational EMF exposure. African countries often face challenges in regulating workplace safety due to a lack of enforcement and limited access to protective equipment. For instance, in many regions, workers do not receive adequate training on safe practices when working near high-voltage lines, which increases the risk of overexposure to EMFs. Moreover, the rapid pace of infrastructure development in Africa means that utility workers often face extended shifts and limited downtime, which may exacerbate the risks associated with prolonged EMF exposure. Implementing stringent occupational health policies and ensuring proper training and protective measures are essential to mitigating these risks. Mobile phone penetration is among the highest in the world, with millions of people using mobile devices for communication, internet access, and financial transactions (GSMA 2020). Mobile phones emit RF radiation, which operates at frequencies between 800 MHz and 2.6 GHz. There has been ongoing debate about whether prolonged mobile phone use increases the risk of brain tumours, particularly glioma and acoustic neuroma. Although the IARC has classified RF radiation as "possibly carcinogenic to humans" (Group 2B), conclusive evidence linking mobile phone use to cancer remains elusive (IARC 2013). In African countries, where mobile phones serve as vital communication tools, it is crucial to raise public awareness about safe mobile phone usage. In rural areas with poor infrastructure, mobile phone base stations are often erected in close proximity to residential areas, raising concerns about chronic RF exposure. Although base stations are designed to emit low levels of RF radiation, continuous exposure over time could pose health risks, particularly for those living near these structures (Ombati et al. 2016). Telecommunication workers who install and maintain mobile phone towers and base stations are at a higher risk of RF exposure compared to the general public. In Africa, this workforce is growing as telecommunication companies expand coverage to underserved regions. However, many workers lack adequate training on the health risks associated with RF exposure and the proper use of protective equipment. In Kenya, a study of telecommunication workers revealed that while workers were aware of RF radiation risks, they did not consistently follow safety protocols, largely due to the lack of enforcement and equipment (Ombati et al. 2016). The study highlighted the need for more stringent occupational health and safety regulations in the telecommunication industry to protect workers from excessive RF exposure.

In the continent, access to MRI technology is still limited, but its use is expanding, particularly in urban hospitals and private healthcare facilities. The increasing availability of MRI machines in African hospitals is a positive development, as it enhances diagnostic capabilities, but it also raises concerns about potential overuse and the adequacy of safety protocols. Patients undergoing MRI scans are exposed to strong magnetic fields and RF pulses, which can induce biological effects such as tissue heating. Although these effects are generally considered safe, repeated exposure to MRI scans over time may have unknown long-term consequences, particularly for vulnerable populations such as pregnant women and children. Ensuring that medical facilities in Africa follow international safety guidelines, such as those set by the International Electrotechnical Commission (IEC), is essential for minimizing risks to patients (IEC 2015). Healthcare workers who operate MRI machines are at risk of exposure to both the static magnetic fields and RF radiation generated during scans. MRI technicians may not have access to the necessary protective equipment or training on safe practices. A study conducted in a South African hospital found that MRI operators experienced dizziness, nausea, and headaches after prolonged exposure to the magnetic fields generated by MRI machines (Rathebe 2022). These symptoms are consistent with what is known as "magnetophosphenes," a phenomenon in which workers perceive light flashes due to the interaction of magnetic fields with the retina (Chow et al. 2014). The study recommended improved safety measures, such as shielding and limiting the amount of time operators spend in close proximity to MRI machines, to protect healthcare workers from the potential risks of NIR exposure.

Electric utility structures, telecommunication devices, and MRI machines are major sources of NIR in Africa. The rapid expansion of infrastructure, mobile networks, and healthcare technologies has increased exposure levels among both the general public and workers. While the health risks associated with NIR remain a subject of debate, it is essential for African countries to adopt robust regulatory frameworks and occupational health policies to mitigate potential harms. Raising public awareness, enforcing safety guidelines, and providing adequate training for workers are critical steps in ensuring the safe use of these technologies.

5.4 REVIEW OF STUDIES ON LONG-TERM EFFECTS AMONG THE GENERAL PUBLIC AND WORKERS' EXPOSURE TO NON-IONIZING RADIATION

Studies on the long-term health effects of such exposure to NIR are still in its nascent stages across much of the continent. One of the few countries in Africa where studies on NIR and its long-term health effects have been conducted is South Africa. In a study by van Deventer et al. (2017), public exposure to RF radiation from mobile phone base stations in Johannesburg was measured, and the study concluded that RF exposure levels in residential areas close to base stations were generally below the limits set by the International Commission on NIR Protection (ICNIRP). However, the study called for further investigation into potential long-term health effects, such as cancer, given the widespread concern and the need for more localized studies to establish health correlations. Another South African study by Mugabe et al. (2016) focused on public perceptions of health risks associated with mobile phone base stations. The study found that many residents living near base stations expressed concerns about long-term health risks, including headaches, sleep disturbances, and the possibility of developing cancer. Although no direct link was established between base station radiation and these health issues, the study highlighted the need for further epidemiological research to understand the long-term effects of RF exposure in Africa.

In Kenya, Ombati et al. (2016) investigated RF radiation levels from mobile phone towers in Nairobi and its potential health effects on nearby residents. The study found that the RF radiation levels were within international safety limits but emphasized the need for continuous monitoring as the number of mobile towers continues to rise. Although the study did not find conclusive evidence of adverse long-term health effects, it acknowledged the gap in long-term data, particularly on cancer incidence and neurological disorders among the general public living near these towers. In Egypt, Abdel-Rassoul et al. (2007) conducted one of the first comprehensive studies on the health effects of ELF-EMF exposure among populations living near high-voltage power lines. The study found that individuals living within 500 m of power lines reported higher incidences of headaches, sleep disturbances, and memory loss compared to those living farther away. Although the study could not conclusively link these symptoms to ELF-EMF exposure, it highlighted the need for further research into potential long-term effects such as cancer and neurodegenerative diseases. Okonofua et al. (2018) explored the effects of ELF-EMF exposure on residents living near power lines in Lagos. Similar to the Egyptian study, residents reported higher incidences of sleep disorders and chronic fatigue. While these symptoms were not definitively linked to ELF-EMF exposure, the study recommended continuous monitoring of public health in areas near power lines and more research on potential links to chronic diseases. In South Africa, healthcare professionals and patients undergoing MRI scans are exposed to NIR. A study by Botha et al. (2019) examined the potential long-term health effects of frequent MRI scans among patients with chronic illnesses. While the study concluded that MRI exposure was generally safe, it recommended cautious use, especially for vulnerable populations such as pregnant women and children, due to the lack of comprehensive long-term data. Additionally, South African healthcare workers who operate MRI machines have reported experiencing symptoms such as dizziness and nausea, raising concerns about occupational exposure to NIR (Rathebe 2022). Although these symptoms are usually transient, the study recommended further investigation into potential long-term effects, particularly given the increasing use of MRI machines across Africa. While African studies on the long-term effects of NIR exposure are limited, available research suggests potential health risks, particularly for populations living near mobile phone base stations and power lines. More comprehensive, long-term studies are needed to understand the health impacts of chronic exposure to NIR in Africa, particularly given the rapid expansion of telecommunications and electrification projects. In addition to research, there is a need for stronger regulatory frameworks and increased public awareness to mitigate potential health risks and ensure the safe use of technologies that emit NIR.

5.5 REVIEW OF STUDIES ON RISK PERCEPTION OF IONIZING RADIATION SOURCES AMONG THE GENERAL PUBLIC AND WORKERS

In African contexts, exposure to UV radiation is a major issue due to the intense sunlight, affecting both the general public and outdoor workers, while RF exposure from mobile devices and telecommunications infrastructure is on the rise. Although the short-term effects of NIR, such as headaches, fatigue, eye irritation, and skin conditions, are relatively mild, they have the potential to impair quality of life and productivity, especially for workers in high-exposure sectors like agriculture and telecommunications. While African studies on NIR are still relatively limited, those that have been conducted show patterns of health effects consistent with findings from other parts of the world. However, there is a need for further research on NIR exposure in African settings, particularly to better understand the short-term and long-term impacts on health and to develop protective measures for both the general public and workers. In the short term, UV radiation can cause skin erythema (sunburn), photokeratitis (inflammation of the cornea), and immune system suppression (Narayanan et al. 2010). A study in South Africa highlighted the prevalence of acute sunburns in populations living in areas with high UV index levels (Kruger et al. 2018). This short-term exposure increases the risk of long-term effects, such as skin cancer and cataracts, but the immediate effects are more prevalent due to the intense sun exposure typical in many African regions. Moreover, short-term exposure to intense UV radiation has been linked to eye problems such as photokeratitis and conjunctivitis (Lucas et al. 2019). A study conducted in Egypt showed that workers, particularly outdoor labourers, had a higher incidence of eye conditions like cataracts due to UV exposure (El Shazly et al. 2016). These short-term effects can significantly reduce quality of life, especially for vulnerable populations like children and the elderly.

The increasing usage of mobile devices across Africa has raised concerns regarding RF radiation exposure. Mobile phone penetration in Africa is among the highest globally, with some reports estimating over 80% mobile phone ownership in countries like Nigeria and South Africa (GSMA 2020). Short-term effects of RF radiation exposure, such as headaches, fatigue, and cognitive disturbances, have been documented (Sivani and Sudarsanam 2012). A study in Nigeria observed that frequent mobile phone users reported higher rates of headaches and sleep disturbances than less frequent users (Adebayo et al. 2017). However, there is still no conclusive evidence regarding the direct link between short-term RF exposure and health impacts. Workers in various sectors, particularly those in agriculture, telecommunications, and healthcare, are more prone to short-term effects of NIR exposure than the general public. These workers are often exposed to NIR from different sources, including sunlight (UV radiation), RF (telecommunication workers), and ELF radiation from electrical equipment.

Many agricultural workers are exposed to significant amounts of UV radiation due to the nature of their outdoor work. A study conducted in Kenya on tea plantation workers showed that a large number of workers experienced sunburns and eye irritation as a result of long hours of exposure to sunlight (Mwangi et al. 2019). Similarly, in South Africa, farmworkers were found to have higher incidences of acute eye conditions such as photokeratitis and conjunctivitis from direct sunlight exposure (Kruger et al. 2018). The short-term effects of NIR exposure for these workers can lead to missed workdays and decreased productivity, presenting a public health concern. Telecommunication workers are exposed to higher levels of RF radiation compared to the general public, especially those working near transmission towers. A study in Egypt examined the short-term health effects of RF exposure among telecommunication workers and found that many workers experienced symptoms such as headaches, dizziness, and sleep disturbances (Abdel-Rassoul et al. 2007). These effects, although transient, can affect workers' efficiency and safety in the workplace. In a Nigerian study, telecommunications engineers reported higher levels of discomfort and concentration issues after prolonged exposure to RF fields while working near base stations (Adetunji et al. 2015).

Healthcare workers, especially those who operate medical devices such as MRI machines, are exposed to ELF and RF radiation. Short-term exposure to ELF in hospital environments has been associated with acute effects such as headache, nausea, and difficulty concentrating (Zamanian and Hardiman, 2005). In a study conducted in South Africa, medical professionals working with MRI equipment reported increased rates of headaches and dizziness after working with machines for long hours (Maree et al. 2020; Rathebe 2022). These findings indicate a need for stricter safety protocols and protective measures for workers in these environments.

Although African studies on NIR are limited compared to the rest of the world, a growing body of research is emerging. Most of the studies focus on UV radiation due to the high levels of sunlight exposure across the continent. However, there is increasing interest in the potential effects of RF radiation due to the rapid expansion of mobile phone networks and telecommunication infrastructure in Africa. For example, a study from Ghana investigated the biological effects of RF exposure from mobile phone base stations and found elevated levels of stress and headaches in individuals living near base stations compared to those living farther away (Awah et al. 2019). Similar findings were reported in Tanzania, where residents near telecommunication towers reported more frequent episodes of fatigue and difficulty sleeping (Mosha et al. 2018). These short-term effects, while mild, could potentially lead to more severe long-term outcomes if not adequately addressed.

A study in Nigeria assessed the sun exposure levels among construction workers and found that over 70% reported experiencing sunburn within the past year (Ogunleye et al. 2020). The study highlighted that these short-term effects could lead to increased absenteeism and reduced productivity, indicating the need for effective health and safety interventions. Also in Nigeria, a study assessed the health effects of RF exposure in schools and found that students frequently using Wi-Fi reported symptoms such as headaches, dizziness, and concentration difficulties (Eze et al. 2020). The short-term health implications of this exposure are concerning, especially considering the increasing reliance on technology in education.

5.6 REVIEW OF STUDIES ON RISK PERCEPTION OF NON-IONIZING RADIATION SOURCES AMONG THE GENERAL PUBLIC AND WORKERS

The professionals and general public awareness and knowledge about NIR significantly influence risk perception. A study conducted in South Africa revealed that many individuals had limited understanding of the types and sources of NIR, particularly RF radiation from mobile phones and base stations (Mothosola and Rathebe 2023). The study found that while participants were generally aware of potential health risks associated with RF radiation, their knowledge was often based on misconceptions and media reports rather than scientific evidence. Similarly, a survey in Nigeria assessed public perception of EMF from mobile phone base stations and found that most respondents were unaware of the technical aspects of radiation and its potential health effects (Odukoya et al. 2019). This lack of knowledge contributed to heightened anxiety and fear regarding exposure, emphasizing the need for educational campaigns to improve public understanding of NIR. Public sentiment towards NIR sources often incorporates emotional responses such as fear and anxiety, which can significantly shape risk perception. A study in Kenya explored how fear of cancer linked to RF exposure influenced public attitudes towards mobile phone usage (Ngoya et al. 2020). Many participants expressed a strong emotional reaction to potential health risks, despite the lack of conclusive evidence linking mobile phone usage to cancer. This aligns with findings from research in South Africa, where anxiety regarding NIR exposure was found to be prevalent, even among individuals with minimal direct exposure (Mothosola and Rathebe 2023).

Workers in various sectors, particularly telecommunications, construction, and healthcare, are frequently exposed to NIR sources. Studies indicate that these workers often possess varying degrees of awareness about the risks associated with their occupational exposure. A study focusing on telecommunications workers in Nigeria found that while many workers recognized potential risks from RF exposure, they felt that safety protocols were insufficient and poorly enforced (Adetunji et al. 2019). This perception of inadequate protection contributed to their concerns about long-term health effects. In the construction industry, workers reported limited knowledge of the risks associated with UV radiation from sunlight. A study in Ghana found that construction workers were often unaware of the potential short-term effects of sun exposure, such as sunburn and eye damage (Abubakar et al. 2020). This gap in knowledge could lead to increased risk-taking behaviours, such as neglecting protective measures. The perception of risk also influences health and safety practices among workers. A study in South Africa assessed the health and safety measures adopted by construction workers exposed to UV radiation. The findings revealed that while workers acknowledged the risks, compliance with protective measures, such as wearing sunscreen and protective eyewear, was low (Matshabane et al. 2019). This non-compliance was attributed to a belief that short-term effects were not significant enough to warrant protective actions, demonstrating a disconnect between risk perception and actual behaviour.

Media representation of NIR has a profound impact on public and worker perceptions. A study conducted in Nigeria highlighted that sensationalized media coverage of health risks associated with mobile phones contributed to increased public fear and misconceptions about RF exposure (Oladosu et al. 2021). Many respondents indicated that media reports influenced their beliefs about NIR, often leading to exaggerated perceptions of risk. This finding underscores the importance of accurate and balanced media representation in shaping public understanding of NIR. Cultural beliefs and social norms also play a crucial role in shaping risk perception. In some African communities, traditional beliefs regarding health and illness can influence how people interpret risks associated with NIR exposure. A study in Uganda found that cultural attitudes towards modern technology, particularly mobile phones, influenced public perception of associated health risks (Biryabarema et al. 2020). Individuals who viewed mobile phones as symbols of modernity and progress were less likely to perceive them as a health risk, highlighting the interplay between culture and risk perception. The risk perception of NIR sources among the general public and workers in Africa is shaped by various factors, including awareness, emotional responses, occupational exposure, media influence, and cultural beliefs. While there is growing concern regarding NIR, many individuals lack accurate knowledge about its sources and associated risks, leading to heightened anxiety and misperceptions. Addressing these issues through education, community engagement, and responsible media reporting is crucial for improving risk perception and ensuring the safety of both the general public and workers.

5.7 FUTURE STUDIES

As the use of NIR sources continues to grow in Africa, there is an urgent need for comprehensive research on both the health effects and the risk perceptions associated with exposure to these sources. Future studies should focus on several key areas to enhance our understanding and inform public health policies effectively. Longitudinal studies are essential to assess the long-term health effects of NIR exposure on both the general public and workers. Such studies can track changes in health outcomes over time, providing valuable insights into chronic conditions that may arise from prolonged exposure to various NIR sources. For instance, researchers could conduct cohort studies among different worker populations (e.g., telecommunications, construction, healthcare) to monitor long-term outcomes such as skin conditions from UV exposure or neurological effects from RF exposure. Short-term exposure studies can help identify immediate health effects of NIR. Experimental designs, including controlled exposure assessments, can isolate specific sources of NIR, such as mobile phones or Wi-Fi routers, and measure acute responses in individuals. This can be particularly relevant for studying the cognitive and psychological impacts of RF radiation among students or office workers who frequently use electronic devices.

Future studies should specifically target vulnerable populations, such as children, pregnant women, and outdoor workers. Research indicates that these groups may have unique susceptibility to NIR exposure due to biological and environmental factors (Sharma et al. 2021). For example, studies focusing on children's exposure to RF from mobile devices could explore developmental impacts and cognitive functioning. Surveys and focus groups can provide qualitative data on the community's understanding of NIR sources and associated health risks. These studies can inform targeted educational campaigns to improve awareness and knowledge about NIR and its health implications. Understanding the psychological factors that influence risk perception is critical. This could involve employing psychological frameworks to assess how perceived risk influences behaviour, compliance with safety measures, and public attitudes towards technologies that emit NIR.

Research on the role of media in shaping public perception of NIR. This includes analysing how different media formats (social media, traditional news) portray NIR risks and the impact of sensationalized reporting on public anxiety. Studies can investigate effective communication strategies that convey accurate information and promote public understanding of NIR without inducing undue fear. Furthermore, studies should evaluate the effectiveness of existing policies and regulations regarding NIR exposure in Africa. This evaluation could inform policy improvements to enhance worker protection and public safety (Afolabi et al. 2020). Barriers to compliance and identify best practices for training workers in protective measures against NIR exposure must be identified. Collaborative studies involving industries, health professionals, and policymakers can foster a comprehensive approach to worker safety (Adetunji et al. 2019). There is a need for the development and validating measurement tools for assessing NIR exposure levels in various settings. Reliable and accessible measurement devices can help quantify exposure in real-time, enabling better risk assessments (Pillay et al. 2021). Such tools could be invaluable for both public health surveillance and individual monitoring.

5.8 CONCLUSION

As NIR continues to permeate various aspects of daily life in Africa, understanding its long and short-term effects, as well as the associated risk perceptions among the general public and workers, is becoming increasingly crucial. Future studies must adopt a multifaceted approach that combines rigorous scientific inquiry with public health considerations to adequately address the growing concerns surrounding NIR exposure. A comprehensive understanding of both the biological effects of NIR and the psychological dimensions of risk perception is important. By integrating data on health outcomes with insights into how communities perceive and react to risks, researchers can provide a more complete picture that informs effective interventions. Focus should be placed on vulnerable populations such as children, outdoor workers, and those in high-exposure occupations. This can help identify unique susceptibility factors and facilitate the development of targeted health and safety guidelines. Engaging communities through participatory research can foster a better understanding of local beliefs and practices regarding NIR. Education campaigns designed from the insights gained can bridge knowledge gaps and address misconceptions, ultimately improving public health outcomes.

REFERENCES

Abdel-Rassoul, G., El-Fateh, O.A., Salem, A.R., Michael, A., Farahat, F., El-Batanouny, M. and Salem, E., 2007. Neurobehavioral effects among inhabitants around mobile phone base stations. *Neurotoxicology*, 28(2), pp.434–440.

Abubakar, I.R., Twumasi, P. and Nkrumah, F., 2020. Knowledge and practices related to sun exposure among construction workers in Ghana. *Journal of Occupational Health*, 62(1), p.e12105.

Adebayo, B., Ojo, O. and Ajayi, O., 2017. Health effects of mobile phone usage in Nigeria. *African Journal of Health Sciences*, 24(1), pp.32–39.

Adetunji, T., Akinwale, L. and Aderibigbe, A., 2015. Occupational exposure to radiofrequency radiation among telecommunication workers in Nigeria. *Journal of Occupational Health*, 58(5), pp.421–427.

Adetunji, T.A., Akinwale, L.M. and Aderibigbe, A., 2019. Perception of occupational hazards among telecommunications workers in Nigeria. *Journal of Health and Pollution*, 9(23), p.190606.

Afolabi, A.A., Oladosu, O. and Adeyemi, A., 2020. Assessment of compliance with safety standards for electromagnetic field exposure in Nigeria. *Journal of Environmental Health Research*, 19(1), pp.45–60.

Ahlbom, A., Green, A., Kheifets, L., Savitz, D., and Swerdlow, A., 2008. Epidemiology of health effects of radiofrequency exposure. *Environmental Health Perspectives,* 116(5), pp.937–942.

Awah, P.K., Tanle, A. and Osei-Tutu, K., 2019. Health effects of radiofrequency exposure from mobile phone base stations in Ghana. *Journal of Environmental Health*, 82(3), pp.25–30.

Biryabarema, D., Namasasu, J. and Mugisha, M., 2020. Cultural perceptions of mobile phone health risks in Uganda. *Global Health Action*, 13(1), p.1710052.

Botha, R.S., Hunter, A. and Pienaar, J., 2019. MRI safety in South Africa: an investigation into the long-term effects of frequent MRI scans. *South African Medical Journal*, 109(5), pp.330–335.

Chow, K.K., Magnani, B. and Mossman, S., 2014. MRI-related headaches in radiologic technologists: prevalence and prevention. *Journal of Magnetic Resonance Imaging*, 39(4), pp.912–917.

Dhlamini, T., Mthethwa, B. and Thwala, M., 2015. Electromagnetic field exposure from high-voltage power lines in South Africa: public health concerns. *South African Journal of Science*, 111(7–8), pp.55–62.

El Shazly, M., Soliman, M. and Afifi, R., 2016. Cataract incidence among outdoor workers exposed to UV radiation in Egypt. *Ophthalmic Epidemiology*, 23(2), pp.105–111.

Eze, A.A., Okeke, O.I. and Nnaji, O.C., 2020. Health effects of radiofrequency radiation from Wi-Fi exposure in schools in Nigeria. *International Journal of Environmental Health Research*, 30(1), pp.33–42.

GSMA., 2020. *The mobile economy sub-Saharan Africa.* Global System for Mobile Communications Association. London, United Kingdom.

Halgamuge MN, Yak SK, Eberhardt JL., 2015. Reduced growth of soybean seedlings after exposure to weak microwave radiation from GSM 900 mobile phone and base station. *Bioelectromagnetics*, 36(2), pp.87–95.

IARC (International Agency for Research on Cancer). 2011. Non-ionizing radiation, part 2: radiofrequency electromagnetic fields. *IARC Monographs Evaluation on the Carcinogenic Risks to Humans,* 102, pp.1–460.

ICNIRP (International Commission on Non-Ionizing Radiation Protection). 2020. ICNIRP guidelines for limiting exposure to electromagnetic fields (100 kHz to 300 GHz). *Health Physics,* 118(5), pp.483–524.

International Agency for Research on Cancer (IARC), 2013. Non-ionizing radiation, Part 2: Radiofrequency electromagnetic fields. *IARC Monographs on the Evaluation of Carcinogenic Risks to Human.* 102, pp.1–460. Lyon, France: IARC.

International Electrotechnical Commission (IEC)., 2015. Medical electrical equipment – Part 2-33: particular requirements for the basic safety and essential performance of magnetic resonance equipment for medical diagnosis. IEC 60601-2-33:2015.

Jiang, Y., Wang, Z. and Zhao, Y., 2020. Longitudinal study on the health effects of electromagnetic fields in workers exposed to radiofrequency radiation. *International Journal of Environmental Research and Public Health*, 17(7), p.2348.

Kabera, S., Mugisha, J. and Munyaneza, M., 2019. Health effects of mobile phone use among students in Uganda. *Journal of Public Health in Africa*, 10(2), p.964.

Kheifets, L., Repacholi, M., Saunders, R. and van Deventer, E., 2010. The sensitivity of children to electromagnetic fields. *Pediatrics,* 116(2), pp. 303–313.

Kheifets, L., Shimkhada, R. and Greenland, S., 2005. Childhood leukemia and power lines: a risk factor in Africa? *Journal of African Oncology*, 7(2), pp.100–109.

Kruger, A., Wright, C. and Melamed, K., 2018. UV exposure and health risks among South African outdoor workers. *Occupational Health Southern Africa*, 24(4), pp.102–110.

Kumar, S., Ali, S. and Pal, S., 2018. Short-term effects of radiofrequency radiation on human health: a review. *International Journal of Health Sciences*, 12(4), pp.17–24.

Lucas, R.M., McMichael, A.J., Armstrong, B.K., Smith, W.T., 2006. Estimating the global disease burden due to ultraviolet radiation exposure. *International Journal of Epidemiology*, 37(3), pp.654–667.

Lucas, R.M., Yazar, S., Young, A.R., Norval, M. and de Gruijl, F.R., 2019. The global burden of UV-induced eye disease. *Photochemical & Photobiological Sciences*, 18(2), pp.1–8.

Mahlangu, T., Ndlovu, R. and Makhaya, K., 2018. Public perception of electromagnetic fields from mobile phone base stations in South Africa. *Environmental Monitoring and Assessment*, 190(5), pp.312.

Mahlangu, T., Ndlovu, R. and Makhaya, K., 2021. Public perception of electromagnetic fields from mobile phone base stations in South Africa. *Environmental Monitoring and Assessment*, 190(5), p.312.

Maree, M., Van Wyk, L. and Potgieter, D., 2020. Non-ionizing radiation in MRI settings and its health effects on healthcare workers in South Africa. *Journal of Radiology in Africa*, 14(3), pp.21–29.

Matshabane, M.G., Dube, P.H. and Ndhlovu, R., 2019. Knowledge, attitudes, and practices regarding UV radiation exposure among construction workers in South Africa. *International Journal of Environmental Research and Public Health*, 16(14), p.2583.

Matsuyama, Y., Sato, H. and Ohtsu, M., 2020. Development of wearable technology for protection against electromagnetic fields. *Journal of Occupational Health*, 62(1), p.e12105.

Mothosola, R.W., 2023. Exposure knowledge and risk perception of radiofrequency electromagnetic fields among environmental health practitioners in Mpumalanga Province (Masters Dissertation, University of Johannesburg).

Mugabe, T., van Deventer, L. and Nel, A., 2016. Public perceptions of mobile phone base stations in Johannesburg. *South African Journal of Science*, 112(3/4), pp.1–5.

Mwangi, G., Karanja, D. and Njoroge, A., 2019. Sun exposure among tea plantation workers in Kenya: effects on skin and eye health. *Journal of Occupational Health*, 61(1), pp.12–18.

Narayanan, D.L., Saladi, R.N. and Fox, J.L., 2010. Ultraviolet radiation and skin cancer. *International Journal of Dermatology*, 49(9), p.978

Ngoya, A.M., Mainga, B.J. and Mwita, E.C., 2020. Risk perception of mobile phone usage and its associated health impacts among students in Kenya. *Health Science Journal*, 14(1), p.1136.

Odukoya, J.O., Oshin, O.A. and Afolabi, T., 2019. Public perception of electromagnetic fields from mobile phone base stations in Nigeria. *International Journal of Environmental Health Research*, 29(3), pp.274–283.

Ogunleye, O.S., Olayinka, O.I. and Abubakar, A.M., 2020. The effect of solar radiation on the health of construction workers in Nigeria. *Journal of Construction in Developing Countries*, 25(2), pp.95–107.

Okonofua, E.E., Ogundele, F.O. and Adeyemo, A., 2018. ELF-EMF exposure from power lines: effects on public health in Nigeria. *Nigerian Journal of Environmental Health*, 12(2), pp.45–52.

Oladosu, O., Adebiyi, A.O. and Bayo, F., 2021. The role of media in shaping public perception of mobile phone radiation risks in Nigeria. *International Journal of Communication*, 15, pp.21–39.

Ombati, R., Karanja, R. and Mwangi, J., 2016. RF exposure and health risks in telecommunication workers: a Kenyan case study. *Kenya Journal of Occupational Health*, 13(1), pp.39–48.

Pillay, A.L., Mkhize, M.A. and Maharaj, V., 2021. Cognitive effects of mobile phone use among students in South Africa. *Journal of Technology in Behavioral Science*, 6(2), pp.153–162.

Rathebe, P.C., 2022a. Perceived safety of MRI units in the two public hospitals within the central region of South Africa: a pilot study among four MR staff. *Journal of Public Health Research*, 11(3), p.22799036221123386.

Rathebe, P.C., 2022b. Subjective symptoms of SMFs and RF energy, and risk perception among staff working with MR scanners within two public hospitals in South Africa. *Electromagnetic Biology and Medicine*, 41(2), pp.152–162.

Rathebe, P.C., 2023, September. Exposure levels of radiofrequency electromagnetic fields from mobile base stations in Mpumalanga province, South Africa. In *Journal of Physics: Conference Series*, 2591(1), pp. 012–035. IOP Publishing.

Roth, D., Garaigordobil, M. and Herrero, F., 2018. Solar ultraviolet radiation exposure among outdoor workers in Valencia, Spain: implications for photoprotection. *Journal of Photochemistry and Photobiology B*, 83, pp.50–58.

Sharma, S., Verma, R. and Raghav, M., 2021. Understanding the vulnerability of children to non-ionizing radiation: a systematic review. *Environmental Science and Pollution Research*, 28(8), pp.9613–9623.

WHO (World Health Organization)., 2014. Electromagnetic fields and public health: extremely low frequency (ELF). Fact sheet No. 322. Available from: https://www.who.int/peh-emf/publications/facts/fs322/en/.

Index

Note: **Bold** page numbers refer to tables and *italic* page numbers refer to figures.

For Product Safety Concerns and Information please contact our EU representative GPSR@taylorandfrancis.com Taylor & Francis Verlag GmbH, Kaufingerstraße 24, 80331 München, Germany